Coronavirus

Invisible

Killer

**Things you did not know about the tiny
virus that knocked the world to its knees.**

JC Spencer

**Presents a unique understanding of
SARS-CoV-2 *and what may be next***

The VIRUS – Learning how to live with it!

In Memory of
James "Jim" Milton Wing
November 20, 1949 – August 15, 2019

If you don't have it – You may have it.
If you have not had it – You may get it.
When you have it – It is NOT a death sentence.
You will survive and thrive with a healthy immune system.
It is more contagious than the flu
But normally less dangerous.
We must learn to live with and
to overcome the Invisible Killer.

The wise physician welcomes
an integrative approach to provide
advanced scientific solutions.

The wise patient welcomes
wisdom from outside the
allopathic pill box to be cured,
not just be treated 'til death.

– JC Spencer

Content

Learn Hidden Virus Facts

Lessons about Covid-19 in Chronological Order

Preface

In the Aftermath will be a New Normal. To survive and thrive, we must get priorities correct.

The aim of this book is to get you to do something. If not to offer solutions – to ask questions. Expect not to be thanked if you do something. You may be rebuked if you do nothing. And if you do much, expect to not be thanked but to be rebuked.

The "New Normal" following covid-19 will be established by all of us involved. We cannot afford to set around wringing our hands, waiting for something to happen. Your purpose is to bless others with whom you are and with what you have. And, we all have much more than what is tangible – more than what we know. You and I can impact our spheres of influence more than you ever dreamed.

The best is yet to come but it will take all the effort we have. A very different battle is going on. Your battle may be filled with pain and puzzles.

The clash between dreaming and reality will mobilize many to become angry and do harm to themselves and others. But a soft answer from your lips can turn away wrath.

We can dream that all will be well. But there are pent-up feelings of anger in the hearts of the oppressed. We are to comfort the oppressed and calm their fears. Some will not listen and will suffer sudden consequences.

The residual effects of lockdown will push many over the edge – over tipping point as emotion takes reign. Emotion, like a bucking bronco, takes them rapidly up and down and they lose self-control. We are to stretch forth a hand to keep them from falling. Bankruptcy will overcome many who are unable to fully recover. We are to help where possible, encourage when we can, comfort, and provide a shoulder on which to cry. Worry, doubt, fear, anxiety, depression, and even suicide will overtake many. We are to continue to encourage and provide another shoulder on which to cry. Prepare for the worse but pray for and expect the best. Massive catastrophic compounding costs will destroy families and cities. Handouts will dry up and the anger filled spirit of arson will strike a match to burn cities and destroy property as communities become tinderboxes to be ignited. Paid arsonists will torch vulnerable cities.

But be of good cheer, you were chosen to make things better, not worse. We are called to show a better way. This is not Armageddon. This is a vast opportunity for us to do better. Don't blame God for coronvirus... remember, the world has told God to leave. Don't be disturbed – the end is not yet. Stay calm.

Yes, we are in for radical change. Parts of the change may be unbelievably good. Problems will worsen as the Fatherless thirst, that only the Love of God shown by others can quench. You may be delighted to see what God will do with coronavirus.

Foreword

The *Invisible Killer* that knocked the world to its knees, holds mysteries and surprises that will shock and enlighten the reader. Close examination of the culture in the petri dish reveals evidence you were not expecting.

As a skilled master magician, the *Invisible Killer* deceives the eyes of the world and projects fear to every corner of earth. In reality, nothing is as it appears. Back stage, accomplices enable this weak enemy of man to appear strong. Spring loaded traps are explained in mind-blowing detail. The traps are simple but deadly.

Today the virus can cross oceans in a single bound. Global threats again catch humans ill prepared and incompetent.

The virus is next to the greatest threat to the survival of the human race. The only greater threat to the human race is man himself.

The purpose for this book is to encourage the reader to be dissatisfied with status quo. To do better than what all the "experts" instructed us to do.

Had the "experts" applied common sense and the knowledge from Ebola, thousands of lives could have been saved from covid-19. Ebola taught us much and we are already learning new facts from covid-19.

Major plagues have killed millions of people in past years. The plagues of the day are warning signs that the next could become the worst plague in human history because they mutate to increasingly become more deceptive. There are currently over 5,000 strains of viruses with more strains continually added.

My desire is to help you learn how to better protect yourself, your family and your immune system. You have an important obligation to protect those within your sphere of influence, and to bless and love others regardless of circumstances. The future is bright to those who open their eyes, discover the truth and become prepared.

The virus is the ultimate parasite - that can become worse with each mutation. The virus can neither support nor propel itself. It cannot eat, secrete or reproduce on its own. The virus is a skilled liar, deceiver and serial killer.

The Virus is an Invisible Killer

Introduction

The coronavirus is different. But all viruses are ultimate liars, deceivers, and potential serial killers. The virus' ability to transfer false communication causes much havoc in the human body. The virus is the ultimate parasite that cannot support or propel itself. It can neither eat, secrete nor reproduce on its own.

I am plagiarizing parts of this Introduction to *Coronavirus INVISIBLE KILLER*. The author should not mind. I wrote these following words about the virus several years ago.

Viruses seize key positions on the surface of the cells. The virus is a key counterfeiter. Some viruses attack and disable their victims with cruel speed while others take years to harm their host. A virus, as in guerrilla warfare, can lie in wait for a more opportune time when your immune defenses are weakened.

The reason coronavirus is more infectious than the flu virus is that it is A CROWN OF THORNS. The thorns snag onto the cell easier than many other viruses. Fortunately, it is less deadly than the flu except for those with a compromised immune system.

When a virus becomes securely attached to the host cell, it implements its evil plan by telling the cell, *"Don't reproduce yourself. Reproduce me. Here is my RNA to make yourself in my image."*

Let us look at a tiny *system*: If you think little tiny *systems* are not important, listen to the words of the Nobel laureate, professor Joshua Lederberg, PhD at Rockefeller University in New York City, *"Our only real competitors for dominion of the planet remain the viruses."*

A single virus is a tiny evil commando that cannot eat, secrete or propel itself. It is unable to reproduce without the aid of a living cell. The virus follows its preprogrammed instructions to reprogram the cells of another organism, making that organism its host.

By reprogramming the host cell, the virus causes that cell to become a traitor, directing it to cease its *designed function* and, instead, to replicate the invader, producing clones of the virus.

Stephen S. Morse, PhD and Robert D. Brown edited two books on the subject, *Emerging Viruses* and *Evolutionary Biology of Viruses*, where they said, "*The virus ... seizes key positions in the host's body and spreads to other hosts-in-waiting at the first opportunity. ... Some viruses attack and disable their victims with cruel speed. ... Other viruses take years to harm their hosts.* (HIV) *can incubate for up to a decade, allowing the deadly agent plenty of time to pass to new hosts before its ill-effects become apparent. Others* [viruses], *such as herpes simplex, coexist so well we're often unaware of their presence.*"

Some viruses mutate quickly. They do this by changes in the structure of their genetic material. In the virus, the RNA is actually changed. An ever-so-slight mutation of the antigens is all that is necessary to nullify all the antibodies that have been resident within a body for years. This enables the virus to sicken you all over again and again and again. Our bodies design a new antibody to go after the newly-*designed* invaders.

Morse and Brown conclude that "*if we don't take steps to monitor and contain their* [viruses'] *continual thrusts, one of their sorties could one day erupt into a global pandemic.*"

1

The Virus
that Changed
US

Change will be different this time. Change will be different for each of us and we cannot stop change. Let us face the change – embrace the change – and observe the positive benefits coming with these changes.

A clear vision of "what works best" is what we need and is what the people are demanding. The patient is educating the and the results are self-evident.

Whole nations came through the pandemic in flying colors – when they applied "what works best" and rejected much of the "expert advice" of noted organizations blinded by prejudices. Blinders restrict the vision from seeing and accepting any cure outside the allopathic box.

The patient was empowered to take charge of his and her own health because allopathic medicine FAILED before the world at the hands of the WHO, CDC, FDA, and NIH through their unwillingness to integrate more natural remedies .

The old phrase is still applicable today: "*An ounce of prevention is worth a pound of cure.*"

Covid-10 killed all those people because the methodology used by the WHO, CDC, FDA, and NIH is archaic. The culture of the medical establishment needs to return to that of the Father of Medicine – Hippocrates – who coined the phrases, "*Do no harm.*" and "*Let medicine be thy food and food be thy medicine.*"

We need to embrace change that demands new technology be integrated with "*what works best*" with few or no side effects. The new technology in Glycoscience diagnostics and application is a pathway that actually heals – cures – instead of endlessly treating the sick. I know that is where the money is and where the people are – but that is not where health is.

2

To Understand the Ebola Mind-set Explains Covid-19

Methodology

of the WHO, CDC, FDA, and NIH

Lessons Not Learned –
Flash Back to Ebola –
Covid-19 Comparisons

When I wrote the book *Ebola Lies* in 2014, I was amazed at the ignorance, neglect, and deceptions used in the handling a BSL-4 pathagen. There are patters in the methodology that unveil mysteries.

It seems that if history has taught us anything, it has taught us that it hasn't taught us anything.

Let us investigate the comparison of covid-19 with Ebola to let you, the reader decide.

Experts in infectious diseases, microbiologists, immunologists and epidemiologists, were alarmed at the procedures and lack of precautions used to protect the US from Ebola. Doctors, aid workers, and military personnel were in serious danger when they used BSL-3 equipment in a BSL-4 situation. Ebola was declared a Biosafety Level 4 (BSL-4) pathogen. Level 4 is the highest virulence designation for infectious agents. Ebola Zaire, the strain that's gone wild, is considered the worst of the strains. The CDC in a 59 page document explains the significance of these 4 levels.

Effective BSL-4 personal protective equipment was flexible, impenetrable material and a high-efficiency particulate air (HEPA) filtration system designed to filter out viruses. There was profound concern for the thousands of US military personnel who were not wearing BSL-4 gear while working with Ebola patients. Pictures confirm that personnel were wearing *Biosafety Level 3* (BSL-3) protective equipment made from Tyvek suits, paper face masks and a face shield that's open at the bottom. Tyvek is tough material but is easily punctured by any sharp object. Should any contaminant splash onto a paper mask, the wearer will most likely breathe the contamination.

Reports indicated that precautions are inadequate. Health workers were hit hard by the virus. Doctors in Spain scrambled to protect individuals at risk after a nurse was infected with Ebola. A UN worker died of Ebola in a German hospital despite "*intensive medical procedures*." The St. George hospital in Leipzig said the 56-year-old man tested positive for Ebola on Oct. 6 and died a week later, prompting Liberia's U.N. peacekeeping mission to put 41 other staff members under "*close medical observation*." This gave rise for concern in Europe.

Summary of Bad Choices that Made the World More Dangerous with Ebola

- The CDC, AGAIN, lost vials of bird flu viruses, anthrax and other highly pathogenic materials. They totally disappeared with no record of where they went. Did these highly dangerous pathogens fall into enemy hands?

- The first Ebola patient's family in Dallas was left in their apartment with bloody sheets for three days. Government officials know that Ebola spreads by contact with bodily fluids.

- Infectious bodily fluids were pressure sprayed into the air and flushed down the Dallas city drain. Some vomit may have been eaten by a dog which can be infected.

- The CDC director said that it would be counterproductive to cut off air travel from the Ebola infected nations.

- US borders were intentionally left wide open. Border patrol agents report that illegal aliens were entering the US with serious infectious diseases. During the first seven months of 2014, it was reported that at least 71 confirmed undocumented immigrants came across our borders from Ebola outbreak nations.

- The FBI Director said that they could not stop Americans who are fighting for ISIS or other terrorist groups from re-entering the country. This is unbelievable since they have declared war on the US.

- The Ebola epidemic in West Africa started in February 2014. The CDC put in place the protocol and training of medical personnel eight months later.

- An Ebola Czar with no medical experience was appointed. He has stated his thoughts for population control especially in Africa and Asia.

- Terrorist or stupid robbers? Bandits stole a cooler bag of Ebola infested blood from a Red Cross courier in Guinea. Was the blood the target or a terrifying surprise for the bandits?

"Ebola Jihad"

Ebola Jihad was declared on America by ISIS and other terrorist organizations. They intended to purposely infect people and fly them into US. Reports were that terrorists had been arrested at the Texas border. Wrong is called right, bitter is called sweet, darkness is called light, lies are called truth and bad choices continue to come in waves. Political correctness and Executive Orders change neither diseases nor the laws of physics.

3 Reasons for bad choices

The three reasons for all bad choices are ignorance, neglect and rebellion. When all three happen at the same time, it is probably political.

Not having the right knowledge, neglecting to act on right knowledge or consciously choosing what is known to be wrong makes right choices impossible.

Many scientists and world leaders are on record espousing population control as the solution to solve the planet's problems. Hmmm?

Population Control – this is no conspiracy theory

I recall a natural disaster in Bangladesh that killed 100,000 people. An official's response still rings in my ears and sends chills up my spine, *"The monsoon was a failure. It only killed 100,000."*

Global population reduction policies are taught in our universities and espoused by globalists. The idea is to reduce the world's population by 2 billion people in any way possible. Some have gone as far as to openly recommend that homosexuals, poor people, blacks and those who are not productive (the elderly and disabled) be targeted.

This Policy is endorsed by the US State Department's Office of Population Affairs (OPA), established in 1975 by Henry Kissinger. This group drafted the Carter administration's *Global 2000* document, which calls for global population reduction and the policy constitutes conscious depopulation projects.

The globalists believe development programs have created a population time bomb. This is their rationalization for reducing the population by 2 billion people. They believe that if there were no population control there will be civil war and greater food shortages because there are just too many people on the planet. The globalists blame themselves for letting people breed like flies, for increasing the survival rate, for extending life and for lowering the death rate. They consider it their own failure for not reducing the birthrate earlier. So drastic action is now required which helps explain the massive push for abortion.

Famous people in high positions of government and finance claim that the "*population crisis may be a greater threat to national security than nuclear annihilation.*"[36]

When hygiene is less than adequate, infection is more rampant. The Ebola outbreak rages in West Africa and experts believe it is destined for Asia where billions of humans are crowded into ever more confining areas.

From the personal experiences which I have had in Africa, India and the Middle-East, I attest to the fact that the hygiene is inadequate. In fact, the conditions are ripe for the largest genocidal event in world history whether perpetrated by war, Ebola, or another pandemic.

Ron Klain, the Ebola Czar at the time, has no medical experience but he has stated that one of his biggest fears is *"overpopulation especially in Asia and Africa that lack the resources to have a healthy, happy life."* Klain added, *"And I think we've got to find a way to make the world work for everyone."*

Some individuals will continue to influence foreign policy based on a genocidal reduction of the world's population. *"We have a network in place of cothinkers in the government,"* said the OPA case officer. *"We keep going, no matter who is in the White House."*

Bio-weapons are a reality and if Ebola were used in a jihad scenario, it should be treated akin to a nuclear attack on the US. A Missouri physician accuses the CDC of dereliction of duty with Ebola. *"It's reactionary, not responsive."* Dr. Gil Mobley is a microbiologist and trauma physician in Springfield, Missouri. He returned from a medical mission trip in Guatemala and was not asked any questions upon re-entry into the country other than whether he had alcohol or cigarettes. When he came through Customs: *"They didn't ask me where I'd been. They didn't ask me if I'd been sick. They didn't ask me if I'd had a fever. And they didn't thermo-scan for my temperature."* This was after the Ebola case in Dallas.

A microbiology expert with 30 years experience, Dr. William Miller, author of the pioneering book, ___The Microcosm Within: Evolution and Extinction in the Hologenome___," said that the establishment is on the wrong side of medical science and is playing Russian roulette by keeping our borders porous. Dr. Miller explained that the plan will inflict a suicidal wound on US.

Were Pogo and Chicken Little both right?
"We have met the enemy and it is us."
"The sky is falling! The sky is falling!"

But wait, the war is not over! Deception will be exposed and truth will set us free.

Why Ebola Became Political

Politicians have stressed that you cannot let a crisis go to waste. Instead of solving the crisis, too often the response is: *"What can we get out of it."* The more mass confusion created, the more the public is willing to yield control to the politicians. A viral pandemic is most effective in creating a domino effect for control.

Statesman and Senator Ted Cruz (R-TX), says the US Ebola policy *"seems to be dictated by politics rather than a common-sense approach to protecting the American public."*

Donald Trump speculated that there's *"something seriously wrong with ...* refusing to cut off flights to countries with active Ebola cases. Trump added, "... *there's something wrong."*

Dr. Ben Cason says that we should fear martial law and most of the politicians and bureaucrats evidence more concern about political power, wealth and their own selfish desires than they do about our country and its citizens.

Bad political strategy uses the uninformed, the ignorant. The uninformed are played for fools. In 2014, I wrote that Ebola or some future virus may be the ultimate crisis they cannot let go to waste. I predicted that it could be the perfect *"act of nature's crisis"* to make the failures of man seem insignificant.

Add chaos to chaos may be part of the plan. In 2014, the CDC actually issued the statement, *"Returning Ebola medical workers should not be quarantined."*

Clearly, our problems cannot be solved politically. Politics is a symptom. We have learned that treating symptoms instead of the problem will cure nothing. The problems in US are spiritual and moral.

Ebola burned out. Was it herd immunity? Covid-19 is less dangerous than Ebola

Ebola was the big virus but it burned out. Covid-19 was another possibility. With only seven genes, the Ebola virus could write scripts to rival a Hollywood murder thriller. The frontal attack was to disable the immune system generals including the macrophage, killer T cells and other white blood cells. To disable the warriors that are in charge of foreign enemies is a masterful military tactic if you are invading a country. Once the generals have been set aside, Ebola can run lawlessly throughout the whole body.

Soldiers are Dismissed and Their Weapons are Taken

Structured glycoproteins on the surface of healthy cells give life to the cell. An army of an estimated 800,000 to a million glycoprotein receptors sites are on the surface of each healthy human cell. These antennae receive and transmit communication and process DNA data. Each cell has an army of these defenders that make up the communication system for the immune system and every function of the body. The only hope against sickness or any virus is a well equipped, well armed, healthy immune system.

Scientists report that the Ebola virus instructs the infected cells by editing the RNA to produce and secrete non-structured glycoproteins into the blood stream. These broken down glycoprotein snippet pieces appear to serve no function other than as decoys to confuse the immune system.

Wounded Warriors Become Totally Defenseless
The fewer glycoproteins on the cell the sicker the cell. The infected cell grows weaker as the glycoprotein structures are dismantled.

Healthy cells fight off infection and communicate the need for hydration and nutrients. When coronavirus attacks the cells, it is different from other viruses. It is not as dangerous as Ebola which immediately begins to dehydrate the body. Covid-19 attacks the lungs to weaken the host. If the person already has a respiratory weakness, that needs to be immediately addressed.

The Glycoprotein warriors have multifunctional benefits beyond the communication for the body. One function is to occupy all the docking stations on the cell so viruses will not be able to attach to the cell in the first place. Covid-19, Ebola and all viruses have a difficult time destroying healthy cells. The weaker cells succumb.

We may not be finished with Ebola.

Mid-July 2020, WHO announced a growing outbreak of Ebola in the Democratic Republic of Congo and warned of an imminent shortage of funds to fight the deadly disease.

Dr. Matshidiso Moeti, the WHO's regional director for Africa, said, *"The current Ebola outbreak is running into headwinds because cases are scattered across remote areas in dense rain forests."*

3

WHAT ARE SMART SUGARS and what do they have to do with coronavirus? EVERYTHING!

Smart Sugars have been my passion for more than two decades. The science is Glycoscience. Smart Sugars are the building blocks of your immune system, your communication system, and the future of medicine and healthcare.

The term, Glycobiology was coined at Oxford University in 1988. Glyco is Greek for sugar and Glycobiology is a branch of biology which deals with sugars. Scientists have identified about twenty-seven bio active sugars found in nature that I call Smart Sugars. Here is a brief look at a few of these Smart Sugars. More information is available in the book, *__Smart Sugars__*.

- **Mannose:** Studies show that mannose has remarkable health benefits, especially involving the immune system, cognitive functions, and cancer. Your god health depends on it. Mannose is cited by the National Library of Medicine in more than 27,000 research references linked to studies.

- **Fucose:** (not to be confused with fructose) Fucose, with major health benefits, may prevent and treat cancer. The National Library of Medicine cites more than 9,700 research references with more than 1,400 linked to cancer studies.

- **Trehalose:** Trehalose has a clean sweet taste that other Smart Sugars do not have. It can be used instead of table sugar. Trehalose is able to protect the integrity of cells against a variety of environmental stresses such as dehydration, heat, cold and oxidation. Trehalose has an important functionality that aids in the proper folding of proteins. More than 6,000 research references are cited by the National Library of Medicine. Trehalose helps hydrate cells. The sugar Trehalose strengthens the cell membrane and has significant benefits in protecting human cells from extreme conditions. Much research is needed especially in relationship to Ebola.

- **Galactose:** Studies reveal significant galactose health benefits: Babies are dependant upon it to build their immune system and help structure the glycoprotein receptor sites. More than 33,000 references cited by the National Library of Medicine.

- **Glucose:** The medical establishment recognizes the basic sugar glucose as very important to human life. However, it appears to be the most harmful in large quantities, especially for diabetics. Glucose is often used as an intravenous drip in hospitals.

Researchers are giving Smart Sugars serious exploration. More physicians are beginning to incorporate these sugars into their practice by asking their patients to eat them. More individuals are ingesting these natural sugars and discovering health benefits. Drug companies are rushing to synthesize these sugars into new drugs. Many thousands of patents related to these super sugars have been issued, especially since 1995.

We are presently aware of a small but growing number of very significant super sugars. Eight of these sugars were presented by Robert K. Murray, MD, Ph.D., and published in multiple editions of <u>Harper's Biochemistry</u>.

Here are some of the sugars that I consider Smart Sugars, part of the Royal Family of Sugars:

Fucose,	Galactose,	Trehalose,
Glucose,	Glucosamine,	Xylose,
Galactosamine,	Sialyllactose,	Arabinose,

Mannose,   Lactose,   Ribose,
Rhamnose,   N-acetylgalactosamine,
N-acetyl-D-N-acetylneuraminic acid
N-acetylglucosamine.

The benefits have become self-evident and scientists have concluded that super sugars have efficacy individually and synergistically that build cells, coat cells, strengthen cells and cell membranes and aid in the cell manufacturing of glycolipids and glyco-proteins.

Glycoscience is the future of medicine. Some 800,000 to one million transmitting sugar antennae coat each healthy human cell resembling little bushy trees. These antennae are constructed from mannose, fucose, glucose, galactose, xylose, N-acetylglucosamine, and variations of other Smart Sugars.

Each antenna is a glycoprotein or glycan. A glycan is a tiny tree-like structure of these different sugars linked together. A glycoprotein is made of sugar and protein links. Without glycans, the cell cannot live. These glycans actually give us life and intelligence. Without these sugars the human cell would have no life.

Become educated in the facts and take action on your newly found knowledge to better your health. The Endowment for Medical Research offers, as a public service, more free educational material. The general public and healthcare professionals are invited to study our websites that could take hundreds of hours to peruse.

Healthcare professionals and the general public have free access to the *Glycoscience whitepaper* at
www.Glycosciencewhitepaper.com

In January 2020, I began documenting my chronological observation of the coronavirus and the life changing sage it has presented. This observation became a series of Lessons and this book.

New Virus Outbreak is Deadly Serious
**Medical Science may not have the answer
but Glycoscience does!**

Glycoscience NEWS Lesson #4 for 2020

1/25/2020 –

As of this morning, reports are that the coronavirus outbreak has reached US, Texas, and possibly a student at Baylor University. Despite containment efforts, cases are already evidenced in 10 countries, including South Korea, Japan, Thailand, Macau, Hong Kong, and Australia.

Associated Press reports that dozens of cities across China are on look-down and the virus may have already affected more than 50 million people. The Associated Press reports Hong Kong declared a high level emergency and closed schools in an attempt to contain the deadly virus. The outbreak started in the central city of Wuhan and has already left 41 people dead.

The US confirmed a second case of coronavirus on Friday (yesterday - 1/24). The CDC announced that more than 60 people in 22 states are being monitored for possible infection. The World Health Organization confirmed this week that the virus has been transmitted person-to-person, but it remains unclear how easy it is to contract from another individual.

What is the novel coronavirus? We know it's new to humans so we don't know much about it. Some say that it showcases viruses' most cunning genetic weapon. Studies indicate that there are thousands of viruses that mutate faster than humans can keep up. The virus is a known threat to the human race.

King Corona is not a beer or a cigar – it's a Crown
The coronavirus gets its name from the protein protrusions that decorate the surface of the virus like a crown. Virologists say that it has become zoonotic, meaning it has the ability to jump between animals and humans and to copy itself like crazy via the next host. The coronaviruses is an RNA virus and has a single strand of genetic material wrapped in tangled protein.

Coronavirus appears to not be as deadly as Ebola and other viruses. The breakout in China may have been due to low quality of eating standards as it originated from a fish and live-animal market in Wuhan, according the WHO.

What we do know is that your immune system is the best defense against any virus. I quote from my Smart Sugar Lesson #71: *"Viruses are the greatest example of evil personified. A virus is worse than a parasite which only feeds on and lives off of another life. A virus is powerless on its own and, like a parasite, attaches itself to another life form. The virus can speak cell language. It communicates with cell signals. Unable to reproduce on its own, the virus signals this message to the host cell, 'Don't reproduce yourself. Reproduce Me! Here is my DNA. Accept it as your own.'*

"When the human host cell is weak, it has no choice but to accept the voice of the enemy. When the human host cell is strong, healthy, and supported with a viable immune system, it automatically and instinctively rejects the virus. The alarmed cell makes at least two immediate responses (cell calls if you will). 1) 'Don't even think about docking your deceiving viral

vessel in this harbor. My receptor sites are loaded and there is no room for you.' 2) By cell-to-cell communication, the endangered cell sends an alert signal to summon the mother of all big eaters, the macrophage.

"The macrophage (Greek for big eater) is a powerful warrior against toxin and infections but in human bodies with poor immune systems, they are hard of hearing and less active. Macrophages are a type of white blood cell that disables and ingests foreign invaders of the body. Macrophages are usually immobile and loosely connected to blood vessel walls. [The macrophages] are both nonspecific defense system for a good immune system and specific defense when needed for a specified purpose, like [killing] this deadly virus.

A virus can dock on a cell surface if the glycoprotein receptor sites are sparse. The cell is healthy when the glycoproteins cover the cell like thick fuzz on a peach. Specific Smart Sugars are not only the building blocks for a good immune system, they construct the Operating System of the body for all communication. That is really important when you need a macrophage."

The bottom line is that we need to cut off bad sugars and foods and drinks that lower the immune system. It is important to switch out your sugar bowl with a truly healthful sweetener.

Coronavirus - King of Sickness and Death
The Coronavirus Crown of Thorns is Designed to Kill

Glycoscience NEWS Lesson #5 for 2020
01/31/2020 –

Stocks fell broadly today as traders tried to assess the potential economic impact of China's fast-spreading coronavirus.

China's National Health Commission confirmed today that there are 9,692 confirmed cases of the coronavirus, with 213 deaths. Yesterday WHO recognized the virus as a global health emergency citing concern that the outbreak continues to spread to other countries with weaker health systems. WHO's designation was made to help the United Nations health agency mobilize financial and political support to contain the outbreak.

The coronavirus is a threat to 1.4 billion people of China. Meanwhile, the whole planet faces an unsurmountable challenge from 5,000 different viruses already in place and over 100 new viruses reported each month. The mutating virus is the most ardent enemy of the human race outside of the human race itself.

The coronavirus is not a beer or a cigar. Corona claims to be king of sickness and death in that its name comes from the protein protrusions that decorate its surface like a crown of thorns. The CDC states that coronavirus was named for the crown-like spikes on its surface. In reality, there are four main sub-groupings of coronaviruses, known as alpha, beta, gamma, and delta.

The coronavirus is an RNA virus and has a single strand of genetic material wrapped in tangled protein. Coronavirus and other viruses will most likely wreak havoc first in developing countries in less sanitary areas.

Human health conditions are most vulnerable where good hygiene is absent. It appears that infected people can begin spreading the virus before they themselves get sick.

The virus has time on its side. Scientists argue if the virus is dead or alive. The virus is a zombie – powerless on its own. It appears dead and can do nothing on its own. The only power it has is given by the host cells. It may lie dormant in the human body for years awaiting a more opportune time when the body's immune system is weaker.

The virus has an army of legions. For a moment, compare the virus with the old Roman Imperial army. The legions formed the Roman army's elite heavy infantry, recruited exclusively from Roman citizens. The virus make up its number by recruiting unhealthy cells to become like them.

Person-to-person transmission is occurring and WHO has declared the virus an international emergency.

When the cell-to-cell communication is operating properly, the macrophage may come alone or summon an army to the battle with the virus. One macrophage can battle and devour 100 bacteria. I do not yet know the troops needed for viral battles but we have seen some amazing battles won.

A virus can dock on a cell surface if the glycoprotein receptor sites are sparse. The cell is healthy when the glycoproteins cover the cell like thick fuzz on a peach. Specific Smart Sugars are not only the building blocks for a good immune system, they construct the Operating System of the body for all communication. That is really important when you need a macrophage.

The virus is the greatest example of evil personified. An evil virus is designed to kill and is worse than a parasite which only feeds on and lives off of another life. A virus is powerless on its own and, like a parasite, attaches itself to another life form. The virus speaks cell language. It communicates with cell signals. Unable to reproduce on its own, the virus signals this message to the host cell, "Don't reproduce yourself. Reproduce Me! Here is my DNA. Accept it as your own."

Glycoscience holds the affective **offensive** battle strategy in fighting the virus. It is evident that drugs are weak **defensive** weapons against the virus. It is vital that we cut off bad sugars and foods and drinks or anything that lowers the immune system. A simple first step is to change the sugar in your sugar bowl and begin taking authority to strengthen your immune system.

Coronavirus' First Killing Outside of China – Death in the Philippines

Glycoscience NEWS Lesson #6 for 2020

Sunday 02-02-2020 – (palindrome date)

Coronavirus cases are soaring beyond China and the WHO has declared an international public health emergency. Today, the first death was reported outside China since the outbreak. The Philippines health officials announced that a 44-year-old Chinese man died Saturday from coronavirus after flying into the country from Wuhan, the Chinese city of 11 million at the center of the outbreak.

Today there are 14,300 confirmed cases around the world, and 305 people have died. All but one of the deaths have been in China. Nearly 60 million people are effectively in lockdown in an attempt to contain the virus.

More than 160 cases in 26 countries outside of China are confirmed. As governments increase concern, the United States, Australia, and New Zealand announced that they will not allow foreign nationals who have traveled from or transited through China to enter. These countries will quarantine their own citizens who have visited China but allow them to re-enter.

Pandemic fears have troubled global stock markets and forced US and global carriers to cancel flights. British Airways and Australia's Qantas announced that they will no longer fly to mainland China. Delta said it will suspend flights between the US and China starting today. Outbreak is reported to have spread to every province and region of China.

China faces a medical shortage as the decision is made to shut down entire cities. Medical testing reports appear to be delayed which raises concern that the outbreak may be worse than reported.

Scientists disagree if the coronavirus is dead or alive but they all agree that it is capable of penetrating and injecting its RNA into a weak human cell to alter the DNA of that cell so as to reproduce itself.

It is vital that the cell-to-cell communication system of the human body operate properly. When the Operating System (OS) of a computer becomes corrupted – the data becomes corrupted.

When we correct our human Operating System (OS) – we can **Correct Flawed Gene Expression***.   *Title of Training

Vaccines Against Viruses Are Not Working
Mind-set for treatment of viruses is flawed

Glycoscience NEWS Lesson #7 for 2020

02-04-2020 –

Another HIV vaccine trial just ended in disappointing failure. Anthony S Fauci, Director of U.S.'s National Institute of Allergy and Infectious Diseases, which conducted the trial, stated, *"An HIV vaccine is essential to end the global pandemic, and we hoped this vaccine candidate would work. Regrettably, it does not."*

The International AIDS Society noted the pressing need for a working vaccine to arrive soon. However, the virus is winning. That need not be! Human ability to invent vaccines is not keeping up with the virus' ability to mutate.

Research conducted within the last few days reveal a striking similarity between coronavirus and the Aids virus. While this new information is alarming, it has not yet been peer reviewed. The report was pre-published on January 31, 2020 and includes work from the Kusuma School of biological sciences, Indian institute of technology, New Delhi, India and Acharya Narendra Dev College, University of Delhi, New Delhi, India.

Their Abstract opens with, *"We are currently witnessing a major epidemic caused by the 2019 novel coronavirus (2019-nCoV). The evolution of 2019-nCoV remains elusive. We found 4 insertions in the spike glycoprotein (S) which are unique to the 2019-nCoV and are not present in other coronaviruses."*

The coronavirus is capable of impregnating another cell with its RNA. This alters the DNA of the host cell. Viruses actually rape weaker cells to reproduce themselves.

To win the war with the virus, the medical mind-set must change. A personal friend and Medical Doctor here in Texas had remarkable results with treating HIV with applied Glycoscience. The authorities descended on him in an attempt to take his medical license and bankrupt him. *Dr. H. Reg(inald) McDaniel*

Victory against the virus on case-by-case battles must be to improve communication integrity between cells. This is accomplished by modulating the immune system. When the Operating System (OS) of the human body functions with clarity, the lies told by the virus fall on deaf ears.

Disclaimer: No medical claims are made or intended. More research is needed and purchases help support neurological research.

> ## To win the war with the virus, the medical mind-set must change.

Yes, We Can Cure Cancer and Aids Today

**Response to the Medical Science Excerpts
from the President's State-of-the-Union Address**

Glycoscience NEWS Lesson #8 for 2020

AN OPEN LETTER TO
PRESIDENT DONALD J. TRUMP

Donald Trump: "*In the 20th century, America saved freedom, **transformed science**. ... Now, we must step boldly and bravely into the next chapter of this great American adventure, and we must create a new standard of living for the 21st century. [Let us] ... unlock the extraordinary promise of America's future.*"

The President addresses viruses:
"*In recent years we have made remarkable progress in the fight against HIV and AIDS. Scientific breakthroughs have brought a once-distant dream within reach. My budget will ask Democrats and Republicans to make the needed commitment to eliminate the HIV epidemic in the United States within 10 years. Together, we will defeat AIDS in America.*"

In my previous Glycoscience Lesson, I stated, "*To win the war with the virus, **the medical mind-set must change**. A Medical Doctor and personal friend here in Texas had remarkable results with treating HIV with applied Glycoscience. Authorities descended on him in an attempt to take his medical license. They put him into bankruptcy.*"

In my book, **To Kill A Rat**, I expose the FDA and explain why it needs to be changed. The FDA does not allow toxin-free solutions be used to treat and cure the American people. FDA-approved drugs MUST BE POISON. The "LD50 level" means that before the drug is approved, it MUST FIRST KILL 50% of the animals in a study (Lethal Dose 50). This is archaic and must be changed.

Senator Ted Cruz drafted a bill to change the FDA – I have included the Ted Cruz Bill as one chapter. Senator Ted Cruz says, *"We need to tear down the barriers blocking a new era of medical innovation, and the primary inhibitor is the government itself. It's past time to unleash a supply-side medical revolution, so that instead of simply caring for people with debilitating diseases, we cure them ... and embrace a culture of innovation."*

Mr. President, applied Glycoscience is the best answer to Cancer, Aids, and all 5,000 known viruses.

Get the FDA and Big Pharma out of the way and let some battle-scarred Pioneers go to work to Make the Health of America Great Again! We can do it during your Administration.

The FDA has claimed ownership of the words, "treat and cure" and requires I make this disclaimer: *"No medical claims are made or intended. More research is needed and purchases help support neurological research."*

The Virus Chaos

Glycoscience NEWS Lesson #9 for 2020

02/06/2020 –

Today China acknowledged that the death toll from the coronavirus outbreak reached 563. Japan reported that 10 passengers aboard the luxury cruise liner, Diamond Princess, were infected and quarantined for at least another 12 days.

Hospitals in Wuhan are in chaos in a struggle to find enough beds for the thousands of newly infected patients. At midnight Wednesday 2/5/2020, China's national health commission released a report that there were 28,018 confirmed cases throughout the country – a rise of 3,694 and the biggest 24-hour rise plus another 24,702 suspected cases.

Agencies reported that 10 new cases involve four people from Japan, 2 from the US, 2 from Canada, 1 from New Zealand, and 1 from Taiwan. The World Health Organization has declared an international public health emergency. Hundreds of experts will gather in Geneva early next week in an attempt to find a way to fight back against the outbreak by speeding up research for new drugs and vaccines.

The WHO's director-general, Tedros Adhanom Ghebreyesus, yesterday 2/5/2020, asked for $675 million to help countries address the expected spread of the virus. He added that if we do not invest in preparedness now the bill will be much higher.

Hu Lishan, a senior official in Wuhan, announced that the 11 million population city is facing a "severe" shortage of beds. Hu said 8,182 patients had been admitted to 28 hospitals that have a total of 8,254 beds, adding that they also had inadequate supplies of equipment and materials.

Citizens trapped in the chaos are using their cell phones to report to the outside world. Another cruise ship off Hong Kong with 3,600 passengers and crew were quarantined for a second day on pending testing after three positive cases onboard.

Hong Kong's Cathay Pacific Airways asked its 27,000 employees to take three weeks of unpaid leave, saying conditions are grave. American Airlines Group and United Airlines said they will suspend flights to and from Hong Kong, leaving no U.S. carriers flying passengers to Asian area.

Chinese stocks took a hit of nearly $700 billion on Monday with many factories closed, cities cut off and travel links are constricting global supply chains. Hyundai Motor will suspend production in South Korea because of a disruption to the supply of parts. This is the first major car maker to halt production outside of China. Global auto makers have already extended factory closures in China in line with government guidelines including Hyundai, Tesla, Ford, PSA Peugeot Citroen, Nissan, and Honda Motor. Plane maker Airbus has prolonged a planned closure of its final assembly plant in Tianjin, China. Taiwan's Foxconn, maker of phones for global vendors including Apple, aims to gradually restart factories in China perhaps next week but uncertainty hangs in the air.

To win the war with the virus, the medical mind set must change. The FDA needs to allow toxin-free cures be available to the public with proven claims.

Victory against the virus will be on its way through immunology, improvement of cellular communication integrity, and modulation of the immune system. Glycoscience holds the answer to the virus chaos.

More Virus Chaos

Glycoscience NEWS Lesson #10 for 2020

Saturday 2-8-2020 –

Today the first American was killed by the coronavirus. Yesterday, a Japanese man died with symptoms consistent to coronavirus as the epidemic advances.

Both men were in their 60s and had compromised immune systems. The Japanese man appeared to have been diagnosed with viral pneumonia before he was infected with the virus.

Many people have died during the last few decades from various diseases but in the latter stages developed pneumonia which received the blame for their death. Unless it is a traumatic event, death is finalized by inflamation caused by virus or bacteria.

This morning, according to authorities, deaths from the coronavirus in mainland China rose by 86 to 722. Report is that the number of outstanding cases stands at 31,774.

Yesterday (2-7-2020), Li Wenliang died. He was the 34 year young doctor who was reprimanded for raising the alarm about the new coronavirus. Dr. Li, via social media, warned of the virus in an online chat room more than five weeks ago. The police made him an example of what happens to those who do not comply with official demands for secrecy. He was summoned by the authorities and forced to sign a statement denouncing his warning as an unfounded and illegal rumor.

Things change quickly. Dr. Li's death sparked an emotional flash point of sorrow and outrage on Chinese social media. An outpouring of praise for him simultaneously was evident with fury for communist authorities who put politics above public safety. Chinese social media, often fickle, was almost unanimous in its grief for Dr. Li, with eulogies posted from all over China, even calling for freedom of speech. Instead of fighting the trend, Beijing used the state media to transform Dr. Li into a loyal soldier aligned with the government's cause.

Yesterday, U.S. Secretary of State Mike Pompeo said the United States would invest up to $100 million to assist China and support coronavirus efforts by the WHO. The United States has sent nearly 17.8 tons of medical supplies to China, including masks, gowns and respirators.

Yesterday, here in Houston, I attempted to purchase some face masks for use in manufacturing – nothing to do with virus protection. Manager of the medical supply store told us that they just had 52 cases delivered in the morning and all were sold a few hours later. Sam's, Costco, Walmart, and other stores in Houston are completely sold out of face masks.

To win the war with the virus, the medical mind-set must change. The FDA needs to allow toxin-free cures be available to the public with proven claims.

Victory against the virus will be on its way through immunology, improvement of cellular communication integrity, and modulation of the immune system. Glycoscience holds the answer to the virus chaos.

Increasing World Chaos with Cronovirus False Death Count
People from many countries in quarantine

Glycoscience NEWS Lesson #11 for 2020

Sunday 02/09/2020 – BEIJING –

The number of reported deaths grew by 97 today to 908 with more outside of the country. China's health ministry said another 3,062 cases had been reported over the previous 24 hours, raising the mainland's total to 40,171. It is evident that the tally continues to grow but the mathematics do not add up. There is growing chaos.

France closes 2 schools as 5 British visitors got the virus at a ski resort. Malaysia, South Korea and Vietnam reported 1 new case each. The mother of 34-year-old physician Dr Li Wenliang, who died last week, is seeking an explanation from authorities who reprimanded her son for warning about the virus. "*My child was summoned by the Wuhan Police Bureau at midnight. He was asked to sign an admonishment notice.*"

Reports of contact contamination are reported. Nine family members were infected recently after sharing a communal hot pot meal at a reunion. A 24-year-old man and his 91-year-old grandmother tested positive for the virus. His parents and other relatives were also found to be infected. The word on the street is that community outbreak is inevitable despite quarantines.

"*It can at most delay the spread of the disease,*" Chuang Shuk-kwan, an official with Hong Kong's health department, said, [Quarantines] "*can at most delay spread of* [coronavirus]."

The WHO advance team traveling today was led by Dr. Bruce Aylward, a veteran of the global fight against the 2014 Ebola outbreak. There is much chaos about government approval if WHO is going or not going to China. Mr. Cui, the Chinese ambassador, said on CBS's "Face the Nation." *"I'm sure that they will be going to China very soon."* He declined to say whether a team of experts from the CDC would also be allowed into China. He suggested instead that American experts could be admitted as part of the WHO or as individuals.

The lockdown of entire cities and more than 50 million people raises question about containment and outbreak in that community. The fact remains that in the middle of all the precaution the situation is very unpredictable.

Today, there have been 36 cases in Hong Kong while new cases were reported in Japan, South Korea, Vietnam, Malaysia, the U.K. and Spain. Spain confirmed its second case in Mallorca, a popular vacation island in the Mediterranean. The first case was a German tourist diagnosed a week ago in the Canary Islands off northwest Africa. More than 360 cases have been confirmed outside mainland China. Many countries are increasing travel restrictions.

To win the war with the virus, the medical mind-set must change. The FDA needs to allow toxin-free cures be available to the public with proven claims.

Victory against the virus will be on its way through immunology, improvement of cellular communication integrity, and modulation of the immune system. Glycoscience holds the answer to the virus chaos.

Disclaimer: No medical claims are made or intended. More research is needed and purchases help support neurological research.

Deaths Caused by Lack of FREE SPEECH!
Looking at the problem through the wrong end of the microscope

Glycoscience NEWS Lesson #12 for 2020

02/10/2020 –

Absence of FREE SPEECH in China has caused the death of many and may have caused the coronavirus crisis because it did not allow public awareness early enough. The Chinese people are blaming the government.

America is the FREE SPEECH country; but, in the medical arena we have a problem. Under pretension of consumer protection, toxin-free treatments and cures are not FDA-approved. The archaic LD50 rule the FDA instituted in 1927 requires death to 50% of the animals in drug studies so it can be given to humans.

Senator Ted Cruz blames our government: *"We need to tear down the barriers blocking a new era of medical innovation, and the primary inhibitor is the government itself. It's past time to unleash a supply-side medical revolution, so that instead of simply caring for people with debilitating diseases, we cure them ... and embrace a culture of innovation."*

To win the war with the virus, the medical mind-set must change. The FDA needs to allow toxin-free cures be available to the public with proven claims.

Research scientists are scrambling to learn more about Coronavirus and what to do about it. The Lancet medical journal has become the Resource Centre for clinical information. China has shared new information this month on behalf of the National

Clinical Research Center for Child Health and Disorders and Pediatric Committee of Medical Association of Chinese People's Liberation Army.

I am quoting and referencing several points for understanding from the paper published: February 07, 2020 in The Lancet. *"Since December, 2019, a pneumonia of unknown cause, which has clinical manifestations similar to severe acute respiratory syndrome,1, 2 originated in Wuhan, China, and has rapidly spread across China to at least 23 countries. By Feb 5, 2020, the number of laboratory-confirmed cases had exceeded 20 000, with more than 400 deaths. About 100 children were affected, with the youngest being 30 h[ours] after birth. A novel virus named 2019 novel coronavirus (2019-nCoV) was considered to be the causative agent of this pneumonia. Neonates are thought to be susceptible to the virus because their immune system is not well developed, which is of great concern to neonatal medical service providers. Paediatricians and neonatologists belonging to the National Clinical Research Center for Child Health and Disorders and Pediatric Committee of Medical Association of Chinese People's Liberation Army have contributed to the control efforts in China. We aim to elicit a contingency plan for the 2019-nCoV outbreak in neonatal intensive care units (NICUs), mainly focused on diagnostic and discharge criteria, treatment, prevention, and control strategies."*

Adults and children with Coronavirus pneumonia infection with mild flu-like symptoms can rapidly develop acute respiratory distress, respiratory failure, multiple organ failure, and death.

Any benefit of antiviral drugs is uncertain. Antimicrobial drugs may only be beneficial against bacterial infection. Use or overuse of antimicrobial drugs should be avoided. Chinese scientists state that for neonates (new born babies) with acute respiratory distress syndrome manifested by complete opacification of lungs, high-dose pulmonary surfactant replacement, nitric oxide inhalation, and high-frequency oscillatory ventilation might be effective. Intravenous glucocorticoids or immunoglobulin could be tried in some difficult cases. Neonates with intracranial infection and convulsion should be handled according to corresponding guidelines. In critically ill neonates, continuous renal replacement and extracorporeal membrane oxygenation could be implemented if necessary.

Scientists in China and the USA are looking at the viral problem through the wrong end of the microscope. Victory against the virus will be on its way through immunology, improvement of cellular communication integrity, and modulation of the immune system. Applied Glycoscience holds the answer to the virus chaos.

References:

The Lancet -
https://doi.org/10.1016/S2352-4642(20)30040-7

> # Scientists are looking at the viral problem through the wrong end of the microscope.

Tiny Virus Devastates Casinos in China – More Than 1,000 humans have died

Glycoscience NEWS Lesson #13 for 2020

02/11/2020 – BEIJING –
The death toll from the coronavirus climbed to a reported 1,016. The gambling center of China (Macau) is devastated. Number of cases rose to more than 42,700.

Yesterday, the World Health Organization (WHO) gave the coronavirus the official name "covid-19." The acronym stands for coronavirus disease 2019, as the new virus was discovered near the end of 2019.

There are 393 COVID-19 cases outside of China in 24 countries. It is the 13th confirmed case in the United States, and the seventh in California.

The WHO has declared the outbreak a global public health emergency. China keeps the city of Wuhan on lock down in an effort to contain the spread of the pneumonia-like disease.

President Xi made his first public appearance after the death of the 34 year-young doctor who suddenly turned hero for speaking out about the deadly coronavirus.

Hong Kong authorities are not expected to enact mask laws as a mask shortage sent people scrambling to form long lines at pharmacies, while residents distrustful of her administration after months of pro-democracy protests staged a run on toilet paper. An official appeal to Hong Kong citizens is to stay at home as much as possible. They are requested to avoid participation in social activities and friends meeting. It was not a compulsory request as Hong Kong is a free society.

This morning more than 100 people in a public housing estate in Hong Kong were evacuated in the early hours after two people in the tower were confirmed to have contracted the coronavirus, the South China Morning Post reported.

To win the war with the virus, the medical mind-set must change. The FDA needs to allow toxin-free cures to be available to the public with proven claims.

Victory against the virus will be through immunology, improvement of cellular communication integrity, and modulation of the immune system. Glycoscience holds the answer to the virus chaos.

Instead of developing a new drug for each of the thousands of known pathogens and hurling the needle toward the dart board, perhaps the family and government dollar can be better invested in immunology by making the human immune system great.

The FDA needs to allow toxin-free cures to be available to the public with proven claims.

Covid-19 Deaths Spike in China 1st Case in Texas - 15th in the US

Glycoscience NEWS Lesson #14 for 2020

02/13/2020 –

Today the CDC reported the 1st coronavirus case (now named covid-19) in Texas and the 15th case in the US. The patient was quarantined at Lackland Air Force Base in San Antonio following evacuation from Wuhan, China. All cases in the US are expected to recover.

China reported a sharp spike in deaths and infections – count reached 1,367, up 254 from yesterday. Confirmed cases of the fast-spreading virus jumped 15,152 to 59,804.

Officials are scrambling with a broadened scope of diagnoses for the outbreak. Doctors' analysis now combines lung imaging instead of waiting for laboratory test results. The viral spread has infected people in 28 countries. Japan joined the Philippines and Hong Kong in deaths outside China. Japan's Health Ministry announced today that 44 more people on a cruise ship quarantined in the port of Yokohama, near Tokyo, have tested positive for covid-19. The ship now has 219 infections among its 3,700 passengers and crew. If the report is accurate for North Korea, there are now 29 countries represented with the virus.

Vietnam's official media reported that a village of 10,000 northwest of the capital, Hanoi, was put in lockdown due to a cluster of 16 cases.

Wuhan, China has replaced its top officials in response to public criticism of how authorities have handled the epidemic. The health system is under extreme pressure from the fast-moving outbreak.

Qingdao Municipal Bureau of Industry and Information Technology has converted two large textile companies to produce face masks to fight the epidemic. The companies have retrofitted their equipment to produce 60,000 face masks per day in the first phase of production.

The government is turning over leaders in the provincial health commission. State media has reported that many others have been expelled from the party for transgressions related to the epidemic. Hospitals are overcrowded and lack medical supplies. Police reprimanded doctors who share information early on for "*spreading rumors.*"

The Chinese government has placed lockdown on more than 60 million people. WHO spokesman Tarik Jasarevic said the agency is seeking more clarity from China on the updates to its case definition and reporting protocol. He stated, "*The jump in cases today reflects the broader definition.*" A WHO advance team only arrived in China this past Monday. Chinese foreign ministry spokesman Geng Shuang said at a daily online briefing that the team is here to "*discuss specific arrangements for the China-WHO joint mission with the Chinese side.*" Geng continued, "*The purpose of the joint mission is that experts of both sides can have in-depth communication on the situation and efforts of prevention and control, and come up with advice for China and other affected countries.*"

It is important to observe the covid-19 virus and thousands of other viruses in our war with these killers. But, if we are to win this war with the virus, the medical mind-set must change. The FDA needs to allow toxin-free cures to be available to the public with proven claims.

Victory against the virus will not come from developing a new drug for each new pathogen. Victory will come through immunology, improvement of cell-to-cell communication integrity, and the modulation of the immune system on an individual basis. Glycoscience holds the answer to the war.

We cannot afford to hurl needles at the dart board. Family and government dollars can best be invested in immunology by making the human immune system more battle ready. This can be accomplished with improved glycosylation of glycans and glycoprotein on the cell surface.

Virus Epidemic Surges Past 1,500 Deaths with 67,192 Reported Infected World-Wide

Links to video footage out of China by brave journalists

Glycoscience NEWS Lesson #15 for 2020

BEIJING Saturday 02/15/2020

Today, the death toll from China's new coronavirus epidemic surged to at least 1,523 after 139 more human lives were lost in the epicenter of Hubei province. The health commission for the province reported 2,641 new cases of the covid-19 strain.

More than 67,000 people have now been infected. In a news conference, National Health Commission official Liang Wannian stated that the government will attempt to contain the spread of the virus in Wuhan, which has been under virtual lockdown for three weeks.

Regardless of the official Beijing Daily newspaper warning that people failing to obey government orders to quarantine would be punished – several free-wheeling "journalists" have posed on social media dozens of video reports. A Mr. Chen posted what appears to be his last video on February 4 before disappearing: https://www.bing.com/videos/search?q=Chen+Qiushi+video&view=detail&mid=D5D0EDB67CAD2A9EAB5DD5D0EDB67CAD2A9EAB5D&FORM=VIRE

Consequentially, Mr. Chen and several other "journalists" are missing. Their family and friends do not know where they are.

Unfiltered streaming videos out of Wuhan reveal striking footage from ground zero where normally any criticism about government authorities is quickly stripped from social media and those responsible punished. The state department deployed hundreds of journalists to reshape appearance of the outbreak.

This and other videos reflect the growing thirst for free speech in China. The revolt against government censorship broke out on social media immediately after the death of Li Wenliang, the Wuhan doctor who had tried to warn of the virus before officials had acknowledged the outbreak.

China's leader, Xi Jinping, said that officials need to *"strengthen the guidance of public opinion."* One truth seeking journalist said,*"I am scared. In front of me is the virus. Behind me is China's legal and administrative power."* After authorities contacted his parents to learn his whereabouts, he pointed to his camera, he appeared desperate and defiant, *"I'm not ... scared of death. You think I'm scared of you, Communist Party?"*

French Health Minister Agnes Buzyn on Saturday (Feb 15) announced the first European covid-19 death. There are at least 46 cases of coronavirus in Europe with 11 cases in France and 16 in Germany. The total number of confirmed cases in Malaysia is 22. Cruise ship MS Westerdam, carrying 1,455 passengers and 802 crew members, was turned away by five countries. The CDC's director told CNN that the agency is preparing for a widespread outbreak of covid-19 across the United States.

To win the war with the virus, a different medical approach must be made. The FDA needs to allow toxin-free cures to be made available to the public with proven claims.

Victory against the virus will not come from developing a new drug for each new pathogen. Victory will come through immunology, improvement of cell-to-cell communication integrity, and the modulation of the immune system on an individual basis. Glycoscience holds the answer to the war.

We cannot continue to just hurl needles at the dart board. Family and government dollars can best be invested in immunology by making the human immune system more battle ready. This can be accomplished with improved glycosylation of glycans and glycoprotein on the cell surface.

Expert Warnings About Coronavirus
Deaths reach 1,770 – Infections now at 70,548
Experts speak out: NIH, Senator Tom Cotton, CDC, WHO, Bill Gates, China

Glycoscience NEWS Lesson #16 for 2020

02/16/2020 – Houston

No one knows what this virus is going to do. It would be foolish to think that it will just go away. So much is unknown. A vaccine is as much of a football as the virus itself in predicting which way it will bounce. Futile attempts to help may compound the problem. The Chinese state media is reporting that penalties up to life imprisonment can be handed out to people who sell face masks or goggles which don't meet national standards.

The National Institutes of Health (NIH) went on record today that the coronavirus (Covid-19) is on the *"verge"* of pandemic if not contained. Dr. Anthony Fauci, director of the National Institute of Allergy and Infectious Diseases at NIH, in an interview on CBS News' *"Face The Nation"* said there are currently 24 countries [outside of China] that have over 500 cases of Covid-19.

The international medical journal Lancet published a study conducted by highly respected epidemiologists from China who demonstrated that several of the original cases did NOT have any contact with the raw food market. So we are not sure where the virus originated. There are too many unanswered questions and China is resisting help from the CDC. Dr. Fauci stated that it really would be helpful if the CDC could be a part of the WHO group on the ground in China.

Senator Tom Cotton, (R) Arkansas, appeared on Sunday Morning Features with Maria Bartiromo where he flatly stated that what we do know about the coronavirus is that it *"Did not start at the Wuhan Animal Market."* He added that China's only bio-safety Level Four Super Lab is just down the road that researches human infectious diseases.

Bill Gates, addressing the American Association for the Advancement of Science meeting in Seattle, said, *"This is a huge challenge, we have always known that the potential for either a naturally caused or intentionally caused pandemic is one of the few things that could disrupt health systems, economies and cause more than 10 million excess deaths."* He reportedly called the virus *"potentially a very bad situation."* As he was speaking, a news report confirmed a person in Egypt tested positive for the disease. The Gates Foundation has funded $100 million to fight the coronavirus and believes that gene editing and artificial intelligence can save the planet.

On Sunday, a spokesman for China's National Health Commission said slowing case numbers nationally shows that China is controlling the outbreak.

Victory against the virus will not come from developing a new drug for each new pathogen. Victory will come through immunology, improvement of cell-to-cell communication integrity, and the modulation of the immune system on an individual basis. Glycoscience holds the answer to the war.

We cannot continue to just hurl needles at the dart board. Family and government dollars can best be invested in immunology by making the human immune system more battle ready. This can be accomplished with improved glycosylation of glycans and glycoprotein on the cell surface.

Finding the SOLUTION
to the Coronavirus

Death toll is ~2,000 – Infections have hit 30 countries
TO KNOW THE ENEMY IS IMPORTANT
BUT TO KNOW THE SOLUTION IS CRITICAL

Glycoscience NEWS Lesson #17 for 2020

02-17-2020 HOUSTON –

People in chaos don't know what to do and normally make the wrong decisions. The coronavirus (aka SARS-CoV-2 and Covid-19) has come to steal, kill, and destroy. To know the enemy and how it functions are important but to know the solution to victory is more critical. Now is not the time to re-arrange the furniture on the Titanic. It's time for action!

The name coronavirus was narrowed by the WHO to Covid-19 to give a standard format to use for future reference including future outbreaks. Chaos reigns even in the name as the Coronavirus Study Group named it "2019-nCoV" and recognized the virus as a sister to severe acute respiratory syndrome coronaviruses (SARS-CoVs) and designated it as severe acute respiratory syndrome coronavirus 2 (SARS-CoV-2). Attempting to make things clear, the initial confusion seems to have been settled until further notice to the SARS-CoV-2 virus which causes the disease Covid-19.

Scientists are focused on nomenclature and detailed study of the problem. We are looking at the situation through the wrong end of the microscope. We are studying the virus and we should be looking at ourselves. We are studying the virus and designing a dart filled with a drug to throw at it. Let us study what the perfect human cell looks like and apply Glycoscience in an attempt to make cells more perfect. A healthy human cell guards the door against all invaders.

The healthy glycans and glycoproteins protect the cells. Glycans and glycoproteins are the gate keepers of the cell. They form the high tech communication shield around all cells in plants, animals, and humans. The glycans and glycoproteins give LIFE to all cells and protect cells from damage. This is why Glycoscience is the future of medicine and why Glycoscience will impact healthcare. A healthy cell has between 800,000 and one million highly active protective radio frequency transponders defending the cell.

To win the war with the virus, a different medical approach must be made. The FDA needs to allow toxin-free solutions to be made available to the public with proven claims.

Victory against the virus will not come from developing a new drug for each new pathogen. Victory will come through immunology, improvement of cell-to-cell communication integrity, and the modulation of the immune system on an individual basis. Glycoscience holds the answer to the war.

We cannot continue to just hurl needles at the dart board. Family and government dollars can best be invested in immunology by making the human immune system more battle ready. This can be accomplished with improved glycosylation of glycans and glycoprotein on the cell surface.

2,100+ Died & 75,000+ Infected

Power of the Virus is Limited
by Your Immune System
Facts You Need to Know Plus List of
30 Countries Now Infected with Coronavirus

Glycoscience NEWS Lesson #18 for 2020

02/21/2020 –

Coronavirus aka SARS-CoV-2 and Covid-19

In spite of China's efforts to control the coronavirus outbreak, it has spread to 29 other countries. Chaos reigns as ignorance abounds. Here is what we know. To quarantine a population seems like the right thing to do but still the innocent suffer. Two Diamond Princess cruise ship passengers have died from coronavirus and a report indicates that 621 cases are confirmed.

China's PR campaign is working, touting a big drop in new cases of the coronavirus as a sign it has contained the epidemic while China's death toll reports 2,118 deaths. Fear grows as deaths increase in 29 other countries. Iran reported two deaths on Wednesday, the first fatalities in the Middle East.

Foreign Minister Wang Yi announced, "*China's forceful action has contained the spread of the virus inside China and also the spread of the virus to other parts of the world.*"

Meanwhile an Update 02/18/2020 showed the total number of cases of infection at 75,767 with deaths at 2,129. Countries now infected include: Japan 7233; South Korea 1041; Singapore 840; Hong Kong 672; Thailand 350; Taiwan 241; Malaysia 220; Germany 160; Vietnam 160; Australia 150; United States 150;

France 121; Macau 100; United Arab Emirates 90; United Kingdom 90; Canada 80; Philippines 31; India 30; Italy 30; Iran 22; Russia 20; Spain 20; Belgium 10; Cambodia 10; Egypt 10; Finland 10; Nepal 10; Sri Lanka 10; Sweden 10.

Chaos onboard cruise quarantine: Japan's government faces criticism over quarantine measures on the Diamond Princess cruise ship. The ship, moored in Yokohama, may be the biggest coronavirus cluster outside the Chinese epicentre. An infectious specialist at Kobe University protested the quarantine as completely inadequate in terms of infection control.

Richard Brennan, regional emergency director at the World Health Organization (WHO), said China was making *"tremendous progress in a short period of time"* but cautioned that it was not over yet.

The virus is human's only real enemy other than ourselves. Scientists are studying the ever-mutating virus and we cannot keep up. It is not that the virus has a brain. It is an evil mechanical blotch of tangled protein with its own RNA code. The virus is relentless in its robotic mission of destruction.

The virus does not eat, secrete , mobilize, or replicate with but one mission – death. It has but one capability with which to accomplish its mission – impregnate its RNA seed into a weak living cell that does not have the strength or ability to resist.

Healthy glycans and glycoproteins protect the cell from viral harm. Glycans and glycoproteins are the gate keepers of the cell. They form the high tech communication shield around all cells in plants, animals, and humans.

Glycoscience is the future of medicine and when applied properly can impact all healthcare. Glycoscience provides the solution through immunology. A healthy cell has between 800 thousand and one million highly active protective radio frequency transponders defending the cell.

Glycoscience holds the answer to the battles and the war. This can be accomplished through improved glycosylation of glycans and glycoproteins on the cell surface.

Report: 2,359 Deaths 77,794 Infected
FACTS we have learned the last few hours

Glycoscience NEWS Lesson #19 for 2020

Coronavirus 02/22/2020 –

WHO confirmed 77,794 people are infected with the coronavirus, with 2,348 deaths in China and 11 deaths outside of China with fears that the virus has spread to countries in Africa. Director General Tedros Adhanom Ghebreyesus met with African officials from Geneva on Saturday to urge them to prepare and confirmed that WHO has shipped more than 30,000 sets of personal protective equipment to six countries in Africa, and is set to soon ship 60,000 more sets to 19 countries.

Iran reports 5th death among the 28 infected with the virus which indicates that the virus is transmitting farther than previously known. The rise in cases outside of China threatens an outbreak that can become a global pandemic.

New Important Facts we have learned.
1) The incubation period is at least 27 days instead of 14 as previously thought. Case in point is a 70-year-old man in China's Hubei Province who was infected with coronavirus, did

not show symptoms for 27 days. People who were quarantined for 14 days and again tested may be spreading the virus.

2) When carriers of the virus are in close proximity with others – they can easily spread the virus. The spread of the virus is more easily transmitted than thought, continues for longer than thought, and because of increased travel is more global. The growth consequences are significantly more protracted.

Close Proximity
Quarantined passengers roamed the ship, dined on filet mignons, attended theater shows together, drank at the crowded bars, and danced the evening away.

Because all viruses can mutate, ALL viruses are enemies of human life. They come to steal, kill, and destroy. They are evil commandos of tangled protein with an RNA code.

The virus does not eat, secrete, mobilize, or replicate with but one mission – death. It has but one capability with which to accomplish its mission – impregnate its RNA seed into a weak living cell that does not have the strength or ability to resist.

Healthy glycans and glycoproteins protect the cell from harm. Glycans and glycoproteins are the gate keepers of the cell. When they are at attention, they keep the enemy out. They literally form a shield around the cell.

Glycoscience is the future of medicine and impacts all healthcare. Glycoscience provides the solution through immunology.

Glycoscience holds the answer to the battles and the war. This can be accomplished through improved glycosylation of glycans and glycoproteins on the cell surface.

Most Alarming Report yet about Coronavirus – It is Asymptomatic!

Glycoscience NEWS Lesson #20 for 2020

02/23/2020 – Coronavirus update:

Death count in **China** has climbed to 2,442. This weekend, **Italy** locked down 10 northern towns of some 50,000 people after 133 people tested positive for Covid-19 and 2 died. The ban is on exits and entries of this affected area of **Europe**. 89 of the 133 cases are in the Lombardy region, 17 in Veneto, 2 in Emilia Romagna, 1 in Piemonte and 2 in Rome. **South Korea** leader calls for unprecedented steps to stop the spread as more cases are experienced. President Moon Jae-in said the outbreak had reached "*a crucial watershed*" and that "*the next few days will be a very important critical moment.*" **Iran** reports 43 cases and 8 deaths from the coronavirus. **Israeli** Prime Minister Benjamin Netanyahu is holding high level meetings in the Health Ministry's emergency situation room to discuss the coronavirus. **Japan** minister apologizes after a woman was released from quarantine who later tested positive.

The tiny virus challenges the $45 billion cruise industry as trips are cancelled and ships are rerouted. This trend will likely continue until there is a reduction of new cases. Norwegian Cruises, Carnival Cruise Lines, and Royal Caribbean Cruises have cancelled nearly 40 cruises and rerouted over 40.

Here is the Most Alarming Report Yet

Reuters reported that a 20-year-old Chinese woman from Wuhan, the epicenter of the coronavirus outbreak, who showed no signs of infection, traveled 400 miles north to Anyang where she infected 5 relatives. **Chinese scientists reported this case on Friday which is new evidence that the virus can be spread asymptomatically, that is without symptoms, providing no subjective evidence of existence.** A follow-up examination tested her positive.

JAMA, (Journal of the American Medical Association), published the case study that offers clues about how the coronavirus is spreading and provides evidence that the virus may be very difficult to stop. Dr William Schaffner, an infectious disease expert at Vanderbilt University Medical Center, who was not involved in the study, stated, *"Scientists have been asking if you can have this infection and not be ill? The answer is apparently, yes."* This basically makes all testing worthless!

One America News Network (OAN) reports that China's president has warned that the coronavirus outbreak will likely get worse before it gets better. President Xi Jinping announced that the peak of the outbreak has not yet been reached. The WHO said that more needs to be done to contain the disease as the coronavirus spreads in China prisons.

Without question, Glycoscience is the future of medicine and already impacts all healthcare but the public and most health professions are not educated in Glycoscience that provides the solution through immunology.

The virus does not eat, secrete, mobilize, or replicate with but one mission – death. It has but one capability with which to accomplish its mission – impregnate its RNA seed into a weak living cell that does not have the strength or ability to resist.

Glycans and glycoproteins protect human cells from harm. Indeed, glycans and glycoproteins are gate keepers of the cell – they keep the enemy out as they form a shield over the cell.

Healthy cells have 800 thousand to 1million highly active protective radio frequency transponders defending the cell.

Glycoscience holds the answer to the battles and the war. This can be accomplished through improved glycosylation of glycans and glycoproteins on the cell surface.

Lesson on How a Tiny Virus has Turned the World Up-side-down
Keep your eyes on Africa!

Glycoscience NEWS Lesson #21 for 2020

02/24/2020 –

Almost overnight, the coronavirus skyrocketed in Italy by more than 200 cases. Stocks plunged Monday, with the Dow closing down 1,031 points. World markets fell sharply and fear developed of a coming economic pandemic as viral cases outside China spread. If the coronavirus are not contained, a severe global recession is certain.

In the days ahead, keep your eyes on Africa. Expectations are strong for a coronavirus breakout in Africa. Former Ethiopian health minister, Tedros Adhanom Ghebreyesus, is director - general of the World Health Organization (WHO). He is racing against time to prevent the outbreak and a worldwide pandemic.

Tedros is seen as the first responder to the world's public health crises. During the last 2 months, nearly 2,500 deaths have been reported on 4 continents. He has hesitated to declare a pandemic but that may be coming to an end, especially if new cases appear in Africa as they have in Japan, South Korea, Italy, and Iran.

Potential Impact if Covid-19 hits Africa.
Even without a single case yet, Africa is already a major victim of the virus. Many of the 58 countries of Africa have received investments of billions of dollars from China. The outbreak is weakening the Chinese economy, threatening world economic growth and cutting the need for oil and metals that are the lifelines of the African nations. African developing countries are more dependent on China than before.

The International Monetary Fund and the African central bankers are starting to sound alarm bells as the catastrophe unfolds. The South African Reserve Bank and the Zambian central bank acknowledge negative effects from the virus. These countries sell to China much of their exports in diamonds and copper. The Chinese economy impacts the globe more than in years past.

Director-General Tedros' job calls for crisis-management and he is holding near-daily press conferences and internal meetings on covid-19, regularly updating the United Nations leadership on the crisis. He has developed direct communication lines with health ministers and leaders in affected countries.

Tedros has the challenge of a delicate diplomatic relationship with the Chinese government because of China's big investments in many of the 58 countries of Africa, including Ethiopia. Tedros has served as foreign minister and a high-ranking member of its former ruling coalition.

While involved in Ethiopian politics, Tedros helped build China's influence in the country through investments in railway, hydro power, infrastructure projects along with Ethiopia's sugar and telecommunication industries. China has invested a reported $24.5 billion into Ethiopia's infrastructure.

Without question, Glycoscience is the future of medicine and already impacts all healthcare but the public and most health professions are not educated in Glycoscience that provides the solution through immunology.

The virus does not eat, secrete, mobilize, or replicate; but, it does have one mission – death. It has one capability with which to accomplish its mission – impregnate its RNA seed into a weak living cell that does not have the strength or ability to resist.

Glycans and glycoproteins protect human cells from harm. Indeed, glycans and glycoproteins are gate keepers of the cell – they keep the enemy out as they form a shield over the cell.

Healthy cells have 800 thousand to 1 million highly active glycans and glycoproteins forming the protective shell.

Glycoscience holds the answer to the battles and the war. This can be accomplished through improved glycosylation of glycans and glycoproteins on the cell surface.

Disclaimer: No medical claims are made or intended. More research is needed and purchases help support neurological research.

Keep your eyes on Africa!
But, we are not prepared.

Glycoscience NEWS Lesson #22 for 2020

02/25/2020 –

Yesterday, I said, *"Keep your eyes on Africa!"*

Today, I received *The Lancet*, perhaps the world's most revered medical journal. **"A modelling study published in *The Lancet* estimates that Egypt, Algeria and South Africa are at the highest risk of importing new coronavirus cases in Africa."**

We (that is US) are not prepared.

It is not a question of if but when the virus hits US according to health authorities as 110 people across 26 states are being monitored for possible cases of the new coronavirus.

Rollout of test kits are in question after the kits produced inconclusive results during verification testing at state and local health departments. Chaos becomes rampant when you add faulty test kits to a longer incubation period and the fact that the virus is asymptomatic – contagious with no symptom.

Today, the World Health Organization (WHO) reports that the daily toll of new cases in China has peaked and plateaued. Chinese officials reported 508 new cases and 71 deaths as of Monday. As the number of cases slow in China, they are increasing around the world. The WHO expressed concern that Europe is not prepared for a major outbreak. Nearly 80,000 people in 37 countries are infected with at least 2,600 deaths. New cases in Italy increases the risk of an outbreak.

Off the coast of West Africa, on the Spanish resort island of Tenerife, is the 4 Star Resort, H10 Costa Adeje Palace. An Italian guest at the resort tested positive for the coronavirus. Tenerife is a Spanish territory and the largest of the Canary Islands. Report is that around 1,000 guests are booked at the hotel which is a popular resort with British tourists. The patient is reported to be a doctor from Lombardy, the area hard hit by the virus. The doctor is in isolation at a hospital on the island, pending the results of a second test to be conducted in Madrid by Spain's National Center of Microbiology.

Spain previously confirmed 2 cases of the virus, both foreigners who were hospitalized on Spanish islands: a German citizen on La Gomera and a Briton on Majorca. Anxiety rises in Europe as new cases mount including reports in Tuscany and Sicily.

Today, it is reported that the Italian government deployed the Army to the red area and that checkpoints had been installed inside the locked-down areas of those towns.

Infections in Iran has prompted fears of a contagion throughout the Middle East. Iran's deputy health minister, Iraj Harirchi, who has spearheaded the country's efforts to contain the coronavirus has tested positive for the virus. State-run news agency reported that Mr. Harirchi had been experiencing weakness and flu-like symptoms on Monday before holding a news briefing, and tested positive for the virus later in the day.

The count of coronavirus cases and deaths continues to rise in Iran. A well known member of Parliament, Mahmoud Sadeghi, an outspoken critic of the country's hard-liners, posted on Twitter that he also has the virus.

Glycoscience holds many answers and is the future of medicine. It already impacts all healthcare but the public and most health professions are not yet educated in Glycoscience that can provide the solution through immunology.

Viruses do not eat, secrete, mobilize, or replicate; but, they have one mission – death. They have one capability – impregnate RNA seed into a weak living cell that has not the strength to resist. The host cell's DNA is changed to produce the virus.

Glycans and glycoproteins protect human cells from harm. Indeed, glycans and glycoproteins are gate keepers of the cell – they keep the enemy out as they form a protective shield over the cell and leaves less or no docking stations for the virus.

Healthy cells have 800 thousand to 1 million highly active glycans and glycoproteins forming the protective shield.

Glycoscience holds the answer to the battles and the war. This can be accomplished through improved glycosylation of glycans and glycoproteins on the cell surface. Glycosylation is the secret weapon against our enemy, the virus.

> # Viruses do not eat, secrete, mobilize, or replicate; but, they have one mission – death.

Coronavirus Has Landed in Nigeria and Washington – First infections in Africa and US

Glycoscience NEWS Lesson #23 for 2020

03/01/2020 –

The coronavirus death toll is officially more than 3,000 with infections nearing 89,000 people with 76 confirmed cases in the US. People have tested positive in 66 countries.

An evil factor has just entered the picture. This may be why some people become contaminated with no known contact with an infected person. I have seen security videos of very sick people intentionally contaminating elevator buttons in public buildings. Innocent people push the button and may become affected. Do not fear but it is time to take more precautions. Taxi drivers have the covid-19 in Bangkok, Taiwan, Singapore, and Japan.

The covid-19 virus has landed in Nigeria and Washington State. The 2nd death is reported in Washington with 5 additional cases, 1 in critical condition. New York has its 1st confirmed case. Two healthcare workers in California test positive.

England reported 2 new cases on Friday. In Eastern Europe, the 3 countries of Belarus, Estonia and Lithuania each reported their first case. Scotland and the Dominican Republic report their first cases, and Ecuador reports more.

The WHO warned, *"This virus has pandemic potential."* New infections are reported across Europe. Italy, the epicenter of coronavirus in Europe with Northern Ireland and Wales experiencing their first cases on Friday. The number of cases in Germany had nearly doubled by Friday afternoon as France reported 20 new cases. Switzerland reported 9 new cases.

The number of people infected in South Korea shot up to 2,337 on Friday. New cases, thousands of miles apart, highlight how quickly the virus is making its way around the globe.

The Netherlands, Denmark, Estonia, Norway and Romania all reported their 1st infection, joining Italy, Austria, Croatia, France, Germany, Greece, North Macedonia, Spain, Sweden and Britain.

Companies are scaling back air travel. The French cosmetics giant L'Oréal has suspended all business travel for its 86,000 employees until the end of March. Nestlé, the Swiss-based food company, said it would suspend all international business trips for its 290,000 workers at least until mid-March.

"We have before us a crisis, an epidemic that is coming. We know that certain countries are already much more affected than us," President Emmanuel Macron of France said during a visit to the Paris hospital where a coronavirus patient died this week.

Viruses cannot eat, secrete, mobilize, or replicate; they have one mission – death. It has the capability to impregnate its RNA into a weak living cell that has not the strength to resist. The DNA of the host cell is changed to produce the virus instead of itself.

DO NO HARM is the Hippocratic Oath no longer widely practiced. Glycoscience holds toxic-free answers to questions that drugs cannot answer. Glycoscience already impacts all healthcare but the public and most health professionals are not yet educated in the science that can provide the solution through immunology.

Glycans and glycoproteins protect human cells from harm. Indeed, glycans and glycoproteins are gate keepers of the cell – they keep the enemy out as they form a protective shield around the cell membrane that leaves less or no docking stations for the virus.

Healthy cells have 800 thousand to one million active glycans and glycoproteins forming the protective shield.

Glycoscience holds the answer to the battles and the war. This can be accomplished through improved glycosylation of glycans and glycoproteins on the cell surface. Glycosylation is the secret weapon against our enemy, the virus.

11 Dead in Washington – 1 Dead in California

Serious Caution and Care – No Time to Panic or Overreact – Time to Learn

Glycoscience NEWS Lesson #24 for 2020

03/04/2020 –

As of today, with more than 93,000 coronavirus infections – more than 3,200 have died.

*"My people perish for lack of knowledge."** Some well meaning people who do not wish to spread fear, say that the coronavirus is no more dangerous than the flu but this is not factual.

The WHO says the coronavirus is much deadlier than seasonal flu but may not spread as easily. Here are the stats: About 0.1% to 0.2% of those who catch the seasonal flu die. About 3.44% of those who tested positive for covid-19 have died. That calculates to 17 to 34 times more deadly than the flu.

Let us observe the viral activity these first days of March. Aggressiveness has just been compounded with discovery of a more aggressive type of the coronavirus. This may change everything and throw out past knowledge of the virus. The character of the tiny beast is its ever expanding mutating and compounding aggression traits.

Medical experts appear surprised that only a tiny percent (2.4%) of children exposed to the virus experience severe symptoms, and an even tinier proportion (0.2%) become critically ill. There have been no deaths reported so far in young children worldwide.

"This is one of the unusual findings and curveballs that this virus keeps throwing at us," said Dr. Frank Esper, a pediatric infectious diseases specialist at Cleveland Clinic Children's, whose research focuses on viral respiratory infections and newly recognized infectious diseases. *"Normal coronaviruses seem to affect children and adults equally, but this one, for whatever reason, certainly skews more to the adult population."*

Today, I reached out to Ardythe Luxion Morrow, PhD, M.Sc, Professor at Cincinnati's Children's Hospital Medical Center. Dr. Ardythe is a friend and one of the most knowledgeable people on the planet about Smart Sugars as they appear in mother's breast milk. She was an important teacher at one of our Glycoscience Medical Conferences. She taught us how these biological sugars build the immune system and protect us from viruses, toxins, and other pathogens. I hope to soon interview her about her latest work.

Viruses cannot eat, secrete, mobilize, or replicate; they have one mission – death. It has the capability to impregnate its RNA into a weak living cell that has not the strength to resist. The DNA of the host cell is changed to produce the virus instead of itself.

DO NO HARM is the Hippocratic Oath no longer widely practiced. Glycoscience holds toxic-free answers to questions that drugs cannot answer. Glycoscience already impacts all healthcare but the public and most health professions are not yet educated in the science that can provide the solution through immunology.

Glycans and glycoproteins protect human cells from harm. Indeed, glycans and glycoproteins are gate keepers of the cell – they keep the enemy out as they form a protective shield around the cell membrane that leaves less or no docking stations for the virus.

Healthy cells have 800 thousand to one million active glycans and glycoproteins forming the protective shield.

Glycoscience holds the answer to the battles and the war. This can be accomplished through improved glycosylation of glycans and glycoproteins on the cell surface. Glycosylation is the secret weapon against our enemy, the virus.

DO NO HARM
is the Hippocratic Oath
no longer widely practiced.

Chaos Reigns in Travel

Report: 3 Cases in Houston Area – 96,762 in 80 Countries – 3,290 Deaths

Glycoscience NEWS Lesson #25 for 2020

03/05/2020 8:24 PM CT –

80 countries with 96,763 infections reported and 3,290 reported deaths. 6 African countries now with infections Nigeria, Egypt, Algeria, Senegal, Morocco, and Tunisia – 163 cases in the USA with 3 Houston area.

Chaos reigns in travel. Cruise ships dock and airlines park planes. On board the Grand Princess, thousands fear exposure to coronavirus after sailing with 62 passengers who had previously voyaged with a man who died from the virus. He tested positive for the coronavirus on Tuesday. The ship is skipping its call to Ensenada on Thursday and instead is sailing directly back to San Francisco. When the MS Westerdam arrives in Juneau, the ship will have been empty of guests for over a month. Holland America Line canceled its sailings in Asia last month, citing concern over ports turning away its cruise ships.

United Airlines said Wednesday that it will reduce passenger-carrying capacity 20% on international routes and 10% in the US. United officials said they will temporarily ground an unspecified number of planes. United's decision to reduce flying came shortly after Germany's Lufthansa announced it would park 150 planes because of falling demand.

United announced the cuts shortly after several airline CEOs met at the White House with President Donald Trump and Vice

President Mike Pence. The administration is seeking the airlines' help in tracing travelers who might have come in contact with people ill with covid-19. The airline CEOs said their companies have stepped up cleaning procedures to help protect passengers and employees from the virus.

History is revealed in this brief chronological overview of **Headlines** during the last few days: **Handshake? No thanks: Coronavirus changes global habits** – Amid coronavirus, the world closes its doors to China – **Global covid-19 cases pass 95,000** – Coronavirus more deadly than flu but containable says WHO's Tedros – **Trump donates quarterly salary to HHS for coronavirus efforts** – Iran frees 54,000 prisoners to combat coronavirus – **New Zealand and Nigeria report first confirmed cases of coronavirus infection** – Americans should prepare for coronavirus spread in U.S., CDC says – **Coronavirus starts to set off some recession alarm bells** – Feb 18 South Korea coronavirus cases jump by half – **Hospital director at coronavirus epicenter dies from the virus** – Japan to trial HIV medications on coronavirus patients – Feb 17 **88 more people test positive for coronavirus on ship off Japan** – Activist who criticized Xi over coronavirus arrested, colleagues say – Feb 16 **44 Americans on cruise ship docked in Japan test positive for coronavirus** – Feb 13 CDC confirms 15th case of coronavirus in US – **Coronavirus gets official name from WHO: covid-19** – Coronavirus infects residents on different floors of apartment building – **Deaths top 1,000; U.S. reports 13th confirmed case** – Feb 10 40,000 coronavirus cases 'tip of the iceberg' – Feb 8 **Deadliest day in outbreak as China records 86 fatalities** – Feb 4 Macau shutters casinos over coronavirus fears – Feb 3 **Hong Kong reports the city's first death related to the new coronavirus** – Feb 1 Trump administration restricts travel from Nigeria and five other countries – Jan 31 **UK and Russia report their first coronavirus cases** – Jan 30 WHO declares China virus an international emergency – **First human-to-human transmission of coronavirus in U.S. documented in Illinois** – Trump creates task force to lead U.S. coronavirus response – **Russia closes land border with China in precaution against coronavirus** – Jan 29 Coronavirus: China cases overtake SARS, as virus spreads worldwide – **Australia, UK to quarantine citizens who were in Wuhan** – Coronavirus death toll tops 100 as infection rate accelerates – **Coronavirus death toll in China rises to 82** – Jan 22 Deaths rise, alarm mounts over China virus – **China confirms Wuhan virus can be spread by humans** – Jan 21 CDC expects more U.S. cases of China coronavirus – Jan 18 **China's Wuhan city confirms 17 novel coronavirus cases** – Jan 17 – Coronavirus: Second death from SARS - like illness in Wuhan, China – **WHO says new coronavirus could spread, warns hospitals worldwide**

Viruses cannot eat, secrete, mobilize, or replicate; they have one mission – death. It has the capability to impregnate its RNA into a weak living cell that has not the strength to resist. The DNA of the host cell is changed to produce the virus instead of itself.

It has the capability to impregnate its RNA into a weak living cell that has not the strength to resist.

USA - 13 States with 241 Infections & 14 Deaths
Covid-19: 101,765 Infections 3,460 Deaths in 97 Countries

Glycoscience NEWS Lesson #26 for 2020

03/06/2020 - 2 :30PM CT –

Reported cases of infection 101,765. Coronavirus cases surge in US. Dr. Tedros Adhanom Ghebreyesus, director general of WHO said, *"This is not a drill. This is not a time for excuses. This is a time for pulling out all the stops."*

FACTS: Coronavirus and possible 5,000 other known viruses are spreading in the US. The battle with the virus is the lifestyle of the future. If we do not adjust, we will be like the frog in a pan of cool water that is slowly heated. Unaware of the ever increasing temperature of the water – the frog is boiled to death without even the thought of jumping out. Scientists say there may be the second, more dangerous coronavirus strain.

Serious concerns: (1) The number of infections may be much larger because so few people are tested. (2) Because the virus is asymptomatic, many infected may not "yet" test positive. And, (3) Fact that those who are declared "recovered" may still carry the virus to be activated when immune resistance is lower.

On the east coast, New York cases have doubled and thousands are under quarantine. On Thursday, New York State had confirmed 22 cases of the new coronavirus with 18 of them in Westchester County. All of these cases were connected. More than 2,770 New Yorkers are in home isolation, according to the city's Department of Health.

On the west coast, Washington and California have been hit. In Washington, 13 people have died from the virus, most from a Seattle-area nursing home. Thursday, in the Seattle area, 2 Microsoft employees were treated for covid-19. In California, 56 people have been treated for the virus, the most of any state.

Think positively. Be an overcomer and allow situations to strengthen instead of weaken you. The outbreak will escalate in some areas and there will be added frustration but we are to realize that chaos and panic will not resolve the problem. Unrest should result in learning and increasing wisdom to know what to do next.

Focus on more healthful habits, eat better foods, strengthen your immune system, exercise more, drink plenty of clean water. Increase your knowledge for the next steps to take. Don't be like the frog.

Viruses cannot eat, secrets, mobilize, or replicate; they have one mission – death. It has the capability to impregnate its RNA into a weak living cell that has not the strength to resist. The DNA of the host cell is changed to produce the virus instead of itself.

Think positively. Be an overcomer and allow situations to strengthen you instead of weaken you.

Coronavirus now in 102 Countries - USA Deaths 19 - Surprising FACTS and Unanswered Questions in today's Lesson

Glycoscience NEWS Lesson #27 for 2020

03/07/2020 –

The Coronavirus has hit 102 countries. USA death toll rose to 19 today with 428 infections. Italy considers drastic measures to combat the spread of infections. An American sailor in Italy tested positive.

Deaths by state are Florida 2, Washington 16, California 1. There have been 3,594 deaths Worldwide with 106,026 infections reported. 9 countries in Africa are affected: Egypt, Algeria, Nigeria, Tunisia, S Africa, Togo, Camaroon, Senegal, Moracco

A case was confirmed for covid-19 after the person had died. Autopsies may reveal cases not previously tested for the virus. Don't panic but know that something is going on that we do not yet understand. Together, let's solve some of the mysteries.

The doomsday professional predictors say that the best-case scenario is that 15 million die and a $2.4 trillion hit comes to global gross domestic product (GDP). I am not a doomsday teacher nor a conspiracy theorist. However, I do observe facts and my daddy taught me when I was but a lad to observe how people do things and look for a better way because there is always a better way and what you observe may be 180 degrees from the better way.

In these brief Lessons, I will discuss, **"What is coronavirus teaching US?"** I assure you that we will dig for knowledge that will benefit you and your family. My philosophy is, **"More good than bad will come out of ANY situation when we let it!"**

Surprising FACTS and Unanswered Questions in today's Lesson: (1) covid-19 has no vaccine at this time and (2) if it did, how beneficial would it be? (3) Several companies are scrambling to make vaccines. (4) Antibiotics is often the knee-jerk "solution" to infections and inflamation and a lot of medical challenges but it is not for viruses. (5) Besides, we (that's US) have become dependant for our antibiotic supply from outside sources. (6) Guess from where – from China. (7) China even supplies antibiotics for our military. (8) Who sold out America?

The word *"antibiotic"* means *"anti life."* (Anti = not) and (biotic = life). Perhaps we should be using "probiotics" = *"pro life"* and *"Smart Sugars"* instead of regular sugar and sweeteners, to help prevent any virus from taking lives. Yes, we may be looking at the problem through the wrong end of the microscope.

It is time for innovations, flexibility, and adaptation for a robust solution that actually works to render the virus threat TO NO AFFECT.

Viruses are robots. We are not sure they are alive or zombies because they cannot eat, secrete, mobilize, or replicate; they have one mission – death to living cells. It has the capability to impregnate its RNA into a weak living cell that has not the strength to resist. The DNA of the host cell is changed to produce the virus instead of itself.

107 Countries Hit – 110,034 Infected – 3,825 Dead –
New FACT may lead to coronavirus answer!

Glycoscience NEWS Lesson #28 for 2020

03/08/2020 – **Researchers are stumped**

Children seem to be protected against the coronavirus. Of the 3,825 deaths, it appears that no child under 10 has died from covid-19. Some children have been infected with mild symptoms. Older people with ailments are the most vulnerable. There is clear evidence that people with a modulated immune system are more resistant to the covid-19 – or any virus for that matter.

Researchers are not sure why children are less susceptible to the coronavirus. This interesting fact makes the virus considerably different to the flu virus which children easily get. Something is unique in children's bodies that protect them from the covid-19 even though children spread germs more than adults do.

A study published in mid-February said that children 9 and under represent about 1% of the coronavirus cases and none from that group had died. The study was from the Chinese Center for Disease Control and Prevention.

Scientists are baffled about how children respond to the various viruses differently. From the more than 110,000 covid-19 infections, there have been no deaths of children. And, from the 32 million cases of flu, 125 children have died this season, according to the US CDC.

A study published in January in the New England Journal of Medicine confirms that children seem to only have mild cases of covid-19. What is different about these children? Their blood is different. Life is in the blood and we need to discover how their blood is different.

Keep your eyes on Africa and South America. The 1st death in Latin America – The Ministry of Health Saturday (yesterday - 3/7) confirmed the covid-19 death of a 64-year-old man from Buenos Aires who had recently returned from Europe. He suffered from kidney failure, diabetes, hypertension and bronchitis before being infected with the virus.

Viruses are zombies -- lifeless robots, unable to eat, secrets, mobilize, or replicate; they have one mission – death to living cells. It has the capability to impregnate its RNA into a weak living cell that has not the strength to resist. The DNA of the host cell is changed to produce the virus instead of itself.

109 Countries Hit – 113,584 Infected – 3,996 Dead – 607 US Cases
Watch - Wait - Learn as More Good Than Bad Comes out of this Situation

Glycoscience NEWS Lesson #29 for 2020

03/09/2020 5:30 PM CT –

Watch, wait, and learn as chaos and panic produce fear that causes people to do stupid things. Common sense is out the window and uncommon sense is hard to find. Fear causes people to operate on emotion instead of logic. Let the coronavirus make us think. Regardless of what happens more good than bad can come out of any situation.

The stock market is in free fall. Saudi Arabia and Russia have started an oil war that dropped crude prices by 1/3 to 1/2. Saudi Arabia plans to boost output by some 10 million barrels per day. Sounds bad doesn't it? No! The stock market will rebound. Saudi Arabia, Russia, OPEC, and China are weakened and America is strengthened while the consumer pays less for gasoline. Let us watch, wait, and learn. What was meant for harm, will result in good.

The coronavirus shows no sign of slowing its spread in the US as confirmed cases surpass 600 across 34 states and DC. Today US death toll rose to 26, as worldwide deaths approach 4,000. The Grand Princess cruise ship docks today with at least 21 infected passengers. The virus is impacting major economies as seen in Italy and South Korea. Major banks cut growth forecasts. China is now projected to have zero demand growth in 2020. Four members of Congress, including our Senator Ted Cruz, are in quarantine. Seattle is under siege.

Comparing covid-19 with the flu virus, the CDC estimates some 34 million people have gotten the flu so far this season and 350,000 have been hospitalized. They tell us that the flu virus has killed 20,000 this season including 136 children. The CDC tells us to calm down by saying, *"Hundreds of thousands of people die of the flu every year."* Well, that's sure comforting! Your health is your personal responsibility.

The good news is that children are overpowering the covid-19. Not one child is reported lost from the coronavirus.

The SECRET TO OVERCOMING the covid-19 lies in THE RESISTANT FACTOR to be found in the blood of these children. Let's discover what it is!!!

Viruses are zombies -- lifeless robots, unable to eat, secrete, mobilize, or replicate; they have one mission – death to living cells. It has the capability to impregnate its RNA into a weak living cell that has not the strength to resist. The DNA of the host cell is changed to produce the virus instead of itself.

118 Countries Hit – 118,096 Infected – 4,262 Dead 808 US Cases – 28 US Died – The World is Changing

Glycoscience NEWS Lesson #30 for 2020
03/10/2020 – 3PM CT

Prepare – Much of the Change Will be Good

Coronavirus **Locksdown Italy** – Walmart and Sam's Club told their workers they will receive up to two weeks of pay if they're required to quarantine by the government and additional pay may be provided for up to 26 weeks for full-time and part-time hourly workers. The new policy is *"unprecedented in uncharted times."* – Airlines around the world sank deeper into crisis Tuesday with cancellation of thousands of flights – The world economy seems to be reconfiguring itself – New York moves to limit gatherings – New Jersey confirms first coronavirus death – Anxiety rises on Capitol Hill over coronavirus – German hospital creates drive-through – Washington State officials scrambling to understand not only how to stop the spread in these facilities but how it appears to be jumping from one place to the next – Anyone with *"underlying conditions"* are endangered.

When the immune system is already weak, ability to fight any pathogen compounds the problem. The CDC and the American Heart Association recommend that cardiac patients take extra precautions as the coronavirus outbreak grows.

Respiratory infections weakens the body and may erode the muscles in the heart. Inflammation can trigger a rupture. Diabetes can increase the risk of atherosclerosis.

The long incubation period for covid-19 was considered 14 days but a new study published Monday in Lancet found an indicator that coronavirus may be contagious for 8 to 37 days.

According to Enid Neptune, a pulmonologist at Johns Hopkins Medicine, *"When there's a great deal of misinformation in the public arena and when there's much that we don't know yet about the virus, this is the time to use your medical contacts."*

All the data collected to-date indicate that covid-19 is much less severe in children. As of 02/11/2020, the Chinese CDC recorded 44,600 confirmed cases, but only 400 involved kids under 9 years old, and none died. Life is in the blood, and the blood of these children holds the mystery to destroying the covid-19.

The SECRET TO OVERCOMING the covid-19 lies in THE RESISTANT FACTOR to be found in the blood of these children. Let's discover what it is!!!

Viruses are zombies -- lifeless robots, unable to eat, secrete, mobilize, or replicate; they have one mission – death to living cells. It has the capability to impregnate its RNA into a weak living cell that has not the strength to resist. The DNA of the host cell is changed to produce the virus instead of itself.

118 Countries Hit – 126,135 Infected 4,630 Dead – 1,311 US Cases – 38 US Died – WHO Declares Pandemic Today

Glycoscience NEWS Lesson #31 for 2020

03/11/2020 – 9 PM CT –

WHO says coronavirus qualifies as a global pandemic! *"Pandemic is not a word to use lightly or carelessly,"* said Dr. Tedros Adhanom Ghebreyesus at a news conference in Geneva today. There is evidence on six continents of sustained transmission of the virus, which has infected more than 126,000 people and killed more than 4,600, so by scientific measures the spread qualifies as a pandemic.

Regional outbreaks of an illness that spreads unexpectedly affects large numbers of people. People in 40 states and Washington D.C. have tested positive and there have been 36 related deaths. The mayor and judge closed the Houston Rodeo today. More than a million students are affected in their studies in several states.

This evening President Trump addressed the nation and suspended all flights from Europe. Travelers from Europe have *"seeded"* the virus here. We may see more celebrities test positive. Tom Hanks and Rita Wilson joined the 125,000+ today.

Medical scientists are telling us that the covid-19 is 10 times more dangerous than the flu virus contrary to the administration's attempt to lower the fear factor. Dr. Anthony S. Fauci, National Institute of Allergy and Infectious Diseases director, said, *"People say, 'Well, the flu does this, the flu does that.' The flu has a mortality of 0.1 percent. This has a mortality of 10 times that."*

Germany's Chancellor Angela Merkel warns that the worst is yet to come and predicted that the coronavirus was likely to infect about two-thirds of the German population. Her top medical adviser, Dr. Lothar Weiler, appeared with her and added, *"we don't know how fast that will happen."* One of Germany's top virologists recently indicated that it could take a year or two, or even longer, to have that many people become infected.

Glycoscience holds the answer to the battles and the war. This can be accomplished through improved glycosylation of glycans and glycoproteins on the cell surface. Glycosylation is the secret weapon against our enemy, the virus.

Glycoscience is the toxic-free answer to questions that drugs cannot answer. Glycoscience already impacts all healthcare but the public and most health professionals are not yet educated in the science that can provide the solution through immunology.

> **Glycoscience already impacts all healthcare but the public and most health professionals are not yet educated in the science that can provide the solution through immunology.**

Glycosylation is the Glycoscience Solution to the Coronavirus
Finding PEACE and HOPE in Chaos – and Faith to Overcome Fear

Glycoscience NEWS Lesson #32 for 2020

03/12/2020 9 PM CT –

More Americans are washing their hands than ever before. More people are taking responsibility for their own health. A lot of positive things will come out of this covid-19. We will, as a people, come through this challenge stronger than when we went into the problem.

Glycosylation is the secret weapon against the covid-19 enemy. Applied Glycoscience can actually accomplish an immune system makeover. Several specific Smart Sugars can improve glycosylation of glycans and glycoproteins on the surface of human cells.

Applied Glycoscience is the toxic-free answer to questions that drugs cannot answer. Glycoscience already impacts all healthcare but the public and most health professionals are not yet educated in the science that can provide the solution through immunology.

The Glycoscience Institute teaches the values of Hippocrates, the Father of Medicine. His value system started with, FIRST DO NO HARM and LET FOOD BE THY MEDICINE AND MEDICINE BE THY FOOD. The Hippocratic Oath is no longer widely practiced and it is time to get back to it.

The U.S. coronavirus outbreak has prompted "irresponsible rhetoric" from people who have downplayed its seriousness. Mike Pence stated that there will be thousands more cases of coronavirus in US. The administration wants to put the health of the American people first.

"This is the most aggressive and comprehensive effort to confront a foreign virus in modern history. I am confident that by counting and continuing to take these tough measures, we will significantly reduce the threat to our citizens, and we will ultimately and expeditiously defeat this virus," Trump told the nation in his Oval Office address.

The world is battling this international outbreak of a new SARS-like coronavirus called covid-19. The virus has claimed over 4,700 lives and infected over 128,000 people around the world. Today the US cases reached over 1,600 with 40 deaths. Yes, this is serious. But, this too shall pass.

The Virus is an Opossum

Glycoscience NEWS Lesson #33 for 2020

03/13/2020 –

The virus is an opossum! I learned about opossums when I was a farm boy in Missouri. An opossum normally has 2 litters a year. However, I met a pregnant mamma opossum carrying a litter on her back and another litter in her pouch. They multiply. I recall whacking opossum with a stick and I thought I had killed it. I ran to tell my daddy. When we arrived where the opossum "died" – it was nowhere to be found. My father said, "*Opossums play dead when they are threatened.*"

We like the word, "recovered" when it comes to covid-19. But I have questioned the reality of the word. I knew the virus plays opossum. It could also be that previous test kits were faulty, adding to the chaos and confusion.

Following weeks in quarantine after having been tested positive for the virus, a man tested negative. He was declared "recovered" and on February 24, he celebrated victory with his 65 neighbors. Three days later, he tested positive for the coronavirus again. He was re-hospitalized and his neighbors were locked down once again. His current condition is unknown. More than 100 cases are reported of Chinese patients who have been released from hospitals as survivors of the virus, to test positive again for what they call a bewildering mysterious illness.

It is a good thing to see US take responsibility for our health. A lot of positive benefits are coming out of this chaos. We will come through this challenge stronger.

Glycosylation is the secret weapon against viral enemies. Applied Glycoscience can actually accomplish an immune system makeover. Several specific Smart Sugars can improve glycosylation of glycans and glycoproteins on the surface of human cells.

Applied Glycoscience is the toxic-free answer to questions that drugs cannot answer. Glycoscience already impacts all healthcare but the public and most health professionals are not yet educated in the science that can provide the solution through immunology.

The Glycoscience Institute teaches the values of Hippocrates, the Father of Medicine. His value system started with, FIRST DO NO HARM and LET FOOD BE THY MEDICINE AND MEDICINE BE THY FOOD. The Hippocratic Oath is no longer widely practiced and it is time to get back to it.

"All hands on deck" will limit the damage and shorten the duration of the virus. Let's learn, improve our immune systems, and live a cleaner lifestyle.

The virus is an opossum – deceptive, bewildering, and mysterious. When it is not infecting its prey or when it is threatened, it pretends to be dead.

Number of Covid-19 Cases May Increase Next Week – This, too, shall pass! – More Good NEWS!

Glycoscience NEWS Lesson #34 for 2020

03/14/2020 –

Houston seems different with the cancellation of the **Houston Livestock and Rodeo** that has not been interrupted in 80 years.

Theaters are sparse and traffic is light. Tomorrow, larger churches will be closed. I was the only customer at a Sprint store. I saw fear on the face of a waitress when I asked if they had towels. She said, "*No! There is a lot of fear out here.*" I silently prayed for her and made an attempt to calm her fears. We need to bring peace to others with an awareness to take additional measures to protect ourselves and others. Relax (that's easy for me to say) and let's make the best of everything. Remember, *"This, too, shall pass!"*

Don't panic when there is a sharp spike in the number of people who test positive for the virus during this coming week. The sharp spike in the number of infections will be the results of the number of new tests made available.

What else can I do? More good NEWS:

In addition to applied Glycoscience to modulate your immune system, consider increasing natural vitamin C. I am told that it is difficult to take too much vitamin C because what is not used will easily flush. Also, vitamin D is needed. So, consult your healthcare professional.

MADRID — Spain is placing its 47 million inhabitants under partial lockdown, as part of an emergency to combat the coronavirus, Prime Minister Pedro Sanchez said Saturday in his address to the nation. Spain follows Italy as the second highest number of infections of any European country, after Italy, from the coronavirus.

Glycosylation is the secret weapon against viral enemies. Applied Glycoscience can actually accomplish an immune system makeover. Several specific Smart Sugars can improve glycosylation of glycans and glycoproteins on the surface of human cells.

Applied Glycoscience is the toxic-free answer to questions that drugs cannot answer. Glycoscience already impacts all healthcare but the public and most health professionals are not yet educated in the science that can provide the solution through immunology.

A – B – C –
D – E – Z

Those are Steps *and* Vitamins

Glycoscience NEWS Lesson #35 for 2020

WOW, there's sure a lot of chaos and confusion out there from conflicting "experts." I just heard a "*medical professional*" on national television put down vitamins and nutrition. There is certainly a lot of knowledge lacking.

Your knowledge, your education about the virus, knowledge about your immune system, and what you should do next will not only get you through the days ahead – you will rise above the chaos and confusion.

First, the bad news. It's likely to get worse before it gets better. Asymptomatic people are driving the spread of coronavirus more than we initially thought. Report from Massachusetts is that many in a cluster of 92 cases were triggered by asymptomatic employees at a biotech conference. Several studies show that people without symptoms are causing substantial amounts of infections.

People perish for lack of knowledge! It will get better and we will be better because of the challenges. More good news: The President declared today a National Day of Prayer. Consult those who know their ABCs.

Natural vitamins A - B - C - D - E - will help you overcome, not just covid-19, but all viruses.

The truth is that (starting with vitamin A) multiple natural vitamins help support the immune system. B_{12} is one of eight essential B vitamins; it plays a role in the brain, nervous system, and red blood cell formation. Natural C is the wonder vitamin. Keep it in your kitchen – it's lemons.

Drink 1 or 2 freshly squeezed lemons daily but NOT sweetened with table sugar or a toxic sugar substitute.

Natural vitamin D is important (it is not really a vitamin – it's a hormone). Vitamin E is extremely beneficial. Study these vitamins and follow common sense. Natural minerals, especially zinc and magnesium, are helpful.

About the Z? – That is for zinc. Okay, I know it's not a vitamin, but then neither was the D. Viruses don't seem to like zinc. They melt at the sign of the Z.

Glycosylation is the secret weapon against viral enemies. Applied Glycoscience can actually accomplish an immune system makeover. Several specific Smart Sugars can improve glycosylation of glycans and glycoproteins on the surface of human cells.

Natural vitamins A - B - C - D - E will help you overcome, not just covid-19, but all viruses.

The Panic is Worse than the Virus

All Washed Hands on Deck – Neighbor Helping Neighbor

Glycoscience NEWS Lesson #36 for 2020

03-16-2020 – HOUSTON

The family next door moved from England about two-and-half years ago. In two weeks, we welcomed them to Texas with the Great Houston Flood via Hurricane Harvey. We weathered that storm together. He and his wife work for British Petroleum and last week they were ordered to work from home because of covid-19. Sunday he contacted me and wanted me to know that he was there to help Karen and me in any way that we needed help. Neighbor helping neighbor is a beautiful thing coming out of this virus threat.

A lot of good will come out of the days ahead – more good than bad. But, today was bad with the largest number of deaths and the largest economic loss. The world is in a world of hurt. All that will shake is being shaken.

The world is under viral attack and coronavirus is but one of over 5,000 different mutations of viruses. The flu kills a lot more people but we are used to it. You know why it's not news? Perhaps two reasons: 1) We have grown accustom to it; and 2) The vaccines don't work that well.

The mind-set of the medical world is to view each new viral mutation with a new drug. We should be looking at the best means possible to build our immune systems to kill any virus.

Meanwhile, we should look for possible economical ways to safely protect the body while killing that which attacks the body. I saw a Houston-area medical doctor on national television mention that we should have more zinc in our bodies. And, another guest on the show scoffed.

In July 2019, the NIH published a paper entitled: *The Role of Zinc in Antiviral Immunity*. The Abstract is filed in the US National Library of Medicine and available: www.PubMed.gov. The review *"summarizes current basic science and clinical evidence examining zinc as a direct antiviral, as well as a stimulant of antiviral immunity."*

Glycosylation is the secret weapon against viral enemies. Applied Glycoscience can actually accomplish an immune system makeover. Several specific Smart Sugars can improve glycosylation of glycans and glycoproteins on the surface of human cells.

Applied Glycoscience is the toxic-free answer to questions that drugs cannot answer. Glycoscience already impacts all healthcare but the public and most health professionals are not yet educated in the science that can provide the solution through immunology.

The Glycoscience Institute teaches the values of **Hippocrates, the Father of Medicine**. His value system started with, **FIRST DO NO HARM** and **LET FOOD BE THY MEDICINE AND MEDICINE BE THY FOOD**. The Hippocratic Oath is no longer widely practiced and it is time to get back to it.

No Storm Lasts Forever
Will Vaccines Work?

Glycoscience NEWS Lesson #37 for 2020

03-17-2020 –

A storm is caused when a clash of cool air hits a lot of hot air. This storm is unnatural; it is huge and fearsome. The whole world is involved.

Darkness, chaos, and unrest are in the storm. It will get darker before it gets brighter. This unusual storm is caused by a tiny virus. This minute enemy will overpower those who think they are strong but are not. Understand that you can be diligent or neglectful to empower or disempower this foe. Do not fear the storm! What the enemy meant for harm can be used for good.

All that can be shifted and shaken will be shifted and shaken. The wind, in its threshing, will blow the chaff away. The storm will shutdown things that need to be shutdown. The storm will destroy the weak who think they are strong. It will thwart giants in groans of agony. The storm will accelerate things that need to be accelerated.

Nations will stand in awe at what remains after the storm. Let us examine a pre-curser storm that, too, went viral. This earlier storm holds an important lesson. It is a weaker virus that has killed multitudes. Its name is, "influenza." Reports would indicate that the flu is a bigger storm to US than the coronavirus. According to the CDC, more than 19 million people have influenza with 180,000 going to the hospital and 10,000 deaths. So, if vaccines work, why is the flu virus still so rampant?

Influenza vaccine effectiveness is evaluated by the CDC each year which they think may lower protection from the virus by 40% to 60%. But, Peter Doshi, Ph.D., a postdoctoral fellow at Johns Hopkins University School of Medicine, disagrees. Doshi cites studies conducted in 2010 that, "*vaccinating between 33 and 100 people resulted in one less case of influenza*." **Doshi concluded that the flu vaccine is no better than a placebo.**

Glycosylation is the secret weapon against viral enemies. Applied Glycoscience can actually accomplish an immune system makeover. Several specific Smart Sugars can improve glycosylation of glycans and glycoproteins on the surface of human cells.

Applied Glycoscience is the toxic-free answer to questions that drugs cannot answer. Glycoscience already impacts all healthcare but the public and most health professionals are not yet educated in the science that can provide the solution through immunology.

Not Ordinary Times
We Will WIN This War with the Chinese Virus

Glycoscience NEWS Lesson #38 for 2020

03-18-2020 –

West Virginia became the 50th state to confirm a coronavirus infection. The reason? No covid-19 test had been conducted in

West Virginia. The number of confirmed infections worldwide nears 220,000 with deaths nearing 10,000. Italy is the out-of-control epicenter for Europe followed by Spain, Germany, and France.

The covid-19 has now sprinkled its infection to 32 African countries. These viral embers await ignition to sweep destruction across Africa. Pray these embers are quenched and that the wind of redemption will blow across Africa. Our team here in Houston wishes to bless Africa and have established a website at http://AfricaBlessesUS.com .

This pandemic is our opportunity to improve our health by not putting toxins into our bodies. We need to clean up our consumption habits and build resilient immune systems. The immune system is our first line of defense. An offensive battle position is to have a well-modulated immune system that will both protect us and kill the viruses.

"Asymptomatic and mildly symptomatic transmissions [are] *major factor*[s] *in transmission for covid-19,"* said Dr. William Schaffner, a professor at Vanderbilt University School of Medicine and longtime adviser to the CDC. *"They're going to be the drivers of spread in the community."*

For years, Chad Eschweiler served as our Director of Research. He introduced me to the work of Michael Osterholm, Research Director at the University of Minnesota Center for Infectious Diseases. Osterholm in recent days has appeared on several networks as a national advisor on the coronavirus. He has warned that this pandemic could go on for as many as 18 months if we don't lock down to contain the spread of the virus. Today President Trump likened the battle with this Chinese Virus to an all out war that we will win.

Glycosylation is the secret weapon against viral enemies. Applied Glycoscience can actually accomplish an immune system makeover. Several specific Smart Sugars can improve glycosylation of glycans and glycoproteins on the surface of human cells.

Applied Glycoscience is the toxic-free answer to questions that drugs cannot answer. Glycoscience already impacts all healthcare but the public and most health professionals are not yet educated in the science that can provide the solution through immunology.

Chicken Little was Right

But the Chinese Virus is not a Death Sentence if You Have a Good Immune System

Glycoscience NEWS Lesson #39 for 2020

03-19-2020 – HOUSTON

Restaurants, theaters, churches, sporting events closed. The world is upside down – the President is treating the "China virus" like the war it is – Nations are on lock down – Governor closes all casinos in Nevada – Governor asks 40 million residents to stay home in California – Governor after Governor declares state of emergency – Governor prepares to free prisoners into the public in New York to protect criminals from the virus – Governor closes bars, restaurants and schools in Texas – Governor orders all non-life-sustaining businesses closed in Pennsylvania – Washington's lieutenant governor leaving for the priesthood.

Fifty states have the infection from the covid-19. The number of confirmed infections worldwide nears a quarter of a million in 160 countries with more than 10,000 deaths. Count of African countries is at 38. Europe is in meltdown expecting hundreds of thousands of deaths.

"This is not going to be over with by the end of the summer." These are the words of Michael Osterholm, Research Director at the University of Minnesota Center for Infectious Diseases.

The GlycoScience Institute in Houston is educating doctors, healthcare professionals, and the general public on the science of Smart Sugars. The educational starting point for many in this New Frontier of Medicine is the *Glycoscience Whitepaper*.

Glycosylation is the secret weapon against viral enemies. Applied Glycoscience can actually accomplish an immune system makeover. Several specific Smart Sugars can improve glycosylation of glycans and glycoproteins on the surface of human cells.

Applied Glycoscience is the toxic-free answer to questions that drugs cannot answer. Glycoscience already impacts all healthcare but the public and most health professionals are not yet educated in the science that can provide the solution through immunology.

Did You Know...?

FACTS you may not know –
This is certainly nothing to sneeze at.

Glycoscience NEWS Lesson #40 for 2020

03-20-2020 – HOUSTON

It is evident that much misinformation is going viral as fast as the virus itself. To compound the issue, some "experts" are combating excellent natural benefits. Saw an "expert" putting down vitamins. Some natural solutions are a lot more beneficial than is understood and some are even life-saving. Eating garlic, fresh onions, lemons, super-foods, sucking on zinc tablets, drinking a tablespoon of colloidal silver water, and doing a saline nose flush may be very beneficial if you don't over-do-it.

We are told that rapidly spreading coronavirus is primarily through "person-to-person" contact. But touching contaminated surfaces is a real risk. When you cough or sneeze on their hand and then touch a keypad, cell phone, elevator buttons, doorknob, or faucet – the virus may remain dangerous for 72 hours.

TWO MOST IMPORTANT THINGS we can do: (1) support the immune system to be **defensive** and **offensive** to expel the virus from your body. I have written extensively about how this happens. And (2) control what you eat and breathe to make sure you are not harming your immune system with smoke, toxins, or carcinogens as they may appear in that which you enjoy.

Some people are damaging their bodies and may someday say, *"If I knew I was going to live this long, I would have taken better care of myself."* Take care of yourself today. The number of cases may double every few days for weeks. Confirmation today is: 275,427 cases – 11,397 deaths – US nearing 20,000 cases with 269 deaths. 165 countries infected (43 in Africa).

Why Italy?

Glycoscience NEWS Lesson #41 for 2020

03/21/2020 -- ROME

Italy had nearly 800 deaths today. Total deaths in Italy will soon be over 5,000. Nearly 54,000 people in Italy have tested positive for coronavirus. But, WHY is nearly one third of the world's covid-19 deaths in Italy?

Italy is such a beautiful country and people but now it is the epicenter of the coronavirus pandemic. We will in the days ahead learn much at Italy's expense. Why? So far, there is no single answer. It appears to be a number of factors working together.

There is a growing aging population. The two leading countries with the largest percentage of people over 65 is Japan and Italy. Japan is in first place with 27% of its population 65 years of age or older. In second place is Italy, with 23% of its population 65 years of age or older.

According to a report from the WHO, the Italian government released numbers revealing the percentage of deaths from covid-19 by age group. These statistics may vary as the number of deaths increase but they were:

6% of deaths: 90+ years old
42% of deaths: 80 – 89 years old
35% of deaths: 70 – 79 years old and
16% of deaths: 60 – 69 years old

Worldwide number of infections is over 307,000 in 170 countries and territories with over 13,000 deaths. The United States is nearing 27,000 confirmed infections with 340 deaths.

I have mentioned for us to keep our eyes on Africa especially. There are now 46 nations or territories in Africa affected.

The GlycoScience Institute in Houston is educating doctors, healthcare professionals, and the general public on the science of Smart Sugars. The educational starting point for many in this New Frontier of Medicine is the *Glycoscience Whitepaper*.

> # We live in a world of over 5,000 types of viruses. They all act much the same with different personalities and dangers.

Going Viral from 1 to 300,000

1st known case ~Nov. 17, 2019 near Wuhan, China – 100,000 cases took weeks – 100,000 to 200,000 took 12 days -- 200,000 to 300,000 took 3 days.

Glycoscience NEWS Lesson #42 for 2020

03/22/2020 - HOUSTON

We understand *"going viral"* and *"viral speed"* better than we did a few weeks ago. **To understand the virus may be a matter of life and death.**

The coronavirus is a lone entangled wad of protein until it snags its crown of thorns upon a weak helpless cell. It's a respiratory virus. It attacks the lungs by docking on a lung cell receptor site. A healthy person should have no fear of harm because healthy cells resist viruses and cry out for help. Help soon arrives, destroys the virus, and carries it out of the body.

An unhealthy human cell is defenseless and does not have the communication strength to signal a call for help. It is this silence of the cell that welcomes the virus to move into position. Once snared, the cell's silence becomes the time bomb, the actual trigger for the virus to penetrate and inject the sickly cell with its seed of death – the virus' RNA.

We live in a world of over 5,000 types of viruses. They all act much the same with different personalities and dangers. A healthy human cell is insulated and protected by a shield of glycans and glycoproteins that tell the virus to "*Go away*." But, there is much more to the message than just, "*Go away*."

The glycans and glycoproteins are transponders and when they are healthy, the body is healthy. 800,000 to 1,000,000 highly functional communication antennae coat the cell. This sugar-coated array of communication technology analyzes, recognizes, verifies, and rejects the virus. Other messages summon the macrophage and killer T-cells to resolve the problem – to destroy, and remove this enemy completely out of the body.

Flatten the Curve –How You Can Help

Glycoscience NEWS Lesson #43 for 2020

03-21-2020 - SOUTH KOREA

"*Flatten the curve,*" is a term for slowing the spread. Lowering the number of infections can save many lives. To flatten the curve means fewer people are getting the virus. How do we do that? Let us observe and learn from various regions about how they have handled or mishandled the coronavirus.

South Korea has "*Flattened the Curve*." The coronavirus is still there and it is still dangerous with the capability of bursting into flame again. But for now, the curve is flattened. South Korea has a performance worth studying and possibly emulating.

Toward the end of February, South Korea was confirming more than 900 cases per day. It was a bumpy plateau with the number of new confirmed cases below 100 for a few days, then spiking to over 150 cases. One of the main ingredients for continuing the decline of cases is to not become complacent.

It is wise to be a little scary but know we will overcome. More caution came when a 17-year-old boy died. That wake up call was that anyone can get coronavirus if their immune system allows it. Even the young can carry the virus and infect others even if they show no symptoms. This young man tested negative according to the Korean CDC, but a subsequent test was positive.

Cell phones are playing a significant role in fighting the virus in South Korean. The wise make sure their phones are disinfected – while text messages and cell alerts notify where groups or individuals are known to be infected. South Korea has undergone the most massive testing of any country in the world by focusing especially on groups where the disease is known to have spread. Korean Air requests that all passengers arriving in Korea must have a mobile phone installed with a self-diagnosis app.

Glycosylation is the secret weapon against viral enemies. Applied Glycoscience can actually accomplish an immune system makeover. Several specific Smart Sugars can improve glycosylation of glycans and glycoproteins on the surface of human cells.

The glycans and glycoproteins are transponders and when they are healthy, the body is healthy. 800,000 to 1,000,000 highly functional communication antennae coat the cell. This sugar-coated array of communication technology analyzes, recognizes, verifies, and rejects the virus. Other messages summon the macrophage and killer T-cells to resolve the problem – to destroy and remove this enemy completely out of the body.

Keeping US from Becoming the Epicenter – Don't Panic, but Think as if YOU HAVE THE VIRUS

Glycoscience NEWS Lesson #44 for 2020

03/24/2020

The Safest Place on Earth
A resident at the scientific research South Pole Station in Antarctica posted a picture online that boasted of their large supply of toilet paper and noted that the South Pole may be the safest place on earth right now. There is no sign of the coronavirus, there is no physical contact with the outside world, and they have plenty of toilet paper. Living in Antarctica is the ideal way to not infect other people. But...

Meanwhile back on the rest of earth, New York City appears to be the epicenter of the outbreak in the United States. With 800 deaths in the nation, New York has more than double the number of deaths than Washington state – nearing 200 as of today.

Make your home the Safest Place on Earth for you
I believe it is time for each of us to act like we have the virus when it comes to being physically close to other people. Let me explain. My father taught me to handle a gun as if it were always loaded even when I knew it wasn't. That taught me to practice

a level of safety that others did not have. This hit home when a friend's son killed himself accidently with a gun that he thought was empty. Apply this safety thinking to this virus pandemic. Don't be fearful but be cautious.

Victory against the virus is available through immunology, improvement of cellular communication integrity, and modulation of the immune system. Glycoscience holds the answer to the virus chaos.

Glycosylation is the secret weapon against viral enemies. Applied Glycoscience can actually accomplish an immune system makeover. Several specific Smart Sugars can improve glycosylation of glycans and glycoproteins on the surface of human cells.

The glycans and glycoproteins are transponders and when they are healthy, the body is healthy. 800,000 to 1,000,000 highly functional communication antennae coat the cell. This sugar-coated array of communication technology analyzes, recognizes, verifies, and rejects the virus. Another message summons the macrophage and killer T-cells to resolve the problem – to destroy and remove this enemy completely out of the body.

Applied Glycoscience is the toxic-free answer to questions that drugs cannot answer. Glycoscience already impacts all healthcare but the public and most health professionals are not yet educated in the science that can provide the solution through immunology.

Don't be fearful but be cautious.

CLEAN HANDS
Stop the Spread

Think this before touching that, "*Am I spreading the virus and who touched that before me?*" – Precaution can calm fears.

Glycoscience NEWS Lesson #45 for 2020

03/25/2020 –

The CDC tells us that 80% of infections are spread by the touch of hands. We obviously need to be careful, so let's knuckle-bump or elbow-bump for a while; but, when this pandemic has passed, let's return to the handshake, hug, or kiss. Let's be wise and do the super hygienic greetings for now. Its a good thing that the pandemic is causing US to think "*clean*" and develop better and cleaner habits.

For many years, I have had the habit of leaving any restroom cleaner than when I entered. That's just something I do instinctively – clean the restroom. Jokingly, I tell people that I got into that habit when I read the sign, "*Clean Restroom.*"

Yesterday a WHO spokeswoman told reporters that acceleration of infections in the US has a very real potential to make US the epicenter of the world for the coronavirus. The next 14 days may tell the story. Worldwide, we are nearing 500,000 coronavirus infections in nearly every country and territory of the world with a confirmed death toll of more than 21,000. The speed of how fast the virus can travel around the world is determined by the mode of transportation its human host takes. I am very optimistic in the middle of the "storm of gloom." We need to stay informed and stay safe from this tiny invader.

Save a LIFE by Leaving the Trail Behind You Cleaner than the Path Before You. All washed hands on deck to do our part in stopping the spread of any virus. Soap and hot water destroys most viruses, including covid-19. The virus is a wad of protein. Soap dissolves the membrane that holds the virus together and can simply be washed off. Use a hand sanitizer when soap and water are unavailable. Alcohol-based hand sanitizer is okay but there are natural non-alcohol sanitizers better for your hands.

The fact that the coronavirus can remain active for 72 hours should inspire us to clean the surface of everything that anyone else may have touched before we touched it. Filthy lucre may really be filthy. So, wash your hands again.

My wife tells me that I habitually touch my face more often than the average – which is 23 times per hour. Don't know if that is a scientific study or a statistic made up on the spot. Unless you are isolated from others, the possibility of your touching and spreading contamination is astronomically high.

Applied Glycoscience is the toxic-free answer to questions that drugs cannot answer. Glycoscience already impacts all healthcare but the public and most health professionals are not yet educated in the science that can provide the solution through immunology.

We Are in a Petri Dish to Analyze Our Culture
1,000+ Killed in US – 47 African Nations Hit – When will it Peak?

Glycoscience NEWS Lesson #46 for 2020

03/26/2020 –

Let us look into the petri dish. What will we see in the petri dish? Everyone looks at US in the petri dish differently. The biologist, the mass media, and the public see our culture through... that's right, their CULTURE. Our culture will be revealed.

A petri dish reveals the change in the culture over a period of time. How we respond to culture change determines the "solution" we use to correct the problem. The biologist and medical scientist want a miracle drug cure designed for the coronavirus. The mass media sees sensational events to drive ratings and is willing to be inaccurate to attain the objective. When, with the powerful microscope, you look into the petri dish, what do you see? Fear? Pretend for a moment the virus sees you looking at it. It should be terrified. You need to look at yourself, your immune system. We are to put fear in the virus – not the other way around.

Is everyone looking through the wrong end of the microscope? We should understand the virus has no power or authority except what your immune system gives it. Coronavirus is a destroyer of the lungs. Anything you are doing to weaken your lungs – you should stop it NOW.

Misleading Information

The number of deaths is misleading if not diabolical. If you take my next statement out of context, I will look like a supporter of euthanasia. I am not! The half million plus infected humans and the tens of thousands of deaths were the most vulnerable and perhaps would have soon died anyway. **Of those infected, only 0.9% had no underlying condition. 6.0% had hypertension. 6.3% had chronic respiratory disease. 7.3% were diabetic. And, 10.5% had cardiovascular disease.** (*China CDC data*)

The new frontier of medicine, Glycoscience, addresses all these conditions because glycans aid cell-to-cell communication is involved in all diseases. I wish the world could have a copy of the *Glycoscience Whitepaper* – well, you can. Ask for it.

Applied Glycoscience is the toxic-free answer to questions that drugs cannot answer. Glycoscience already impacts all healthcare but the public and most health professionals are not yet educated in the science that can provide the solution through immunology.

To win the war with the virus, the culture of the medical mind-set must change. The FDA needs to allow toxin-free cures to be available to the public with proven claims.

Victory against the virus will be on its way through immunology, improvement of cellular communication integrity, and modulation of the immune system. Glycoscience holds the answer to the virus chaos.

Pandemic Model Changed

China/Italy Connection explained –
NYC now epicenter – Africa is a tinderbox

Glycoscience NEWS Lesson #47 for 2020

03/27/2020 –

Re-calibration of Pandemic Model – We have learned from the coronavirus pandemic that things can change quickly. Fear swept the planet when British scientist Neil Ferguson published a catastrophic forecast from Imperial College predicting 2.2 million Americans and more than 500,000 British would be killed by the new virus. His report was taken seriously by leadership in both countries. This week, Neil Ferguson re-calibrated his pandemic model in testimony before the U.K.'s parliamentary select committee on science and technology stating that he now predicts U.K. deaths from the disease will not exceed 20,000 and could be much lower.

China/Italy Connection – Thousands of Chinese living in Northern Italy are from the Wuhan area where the breakout occurred. Many flights directly from Wuhan to Italy undoubtedly carried the virus.

We (that's US) have reportedly bypassed China
Confirmed are ~105,000 cases with over 1,700 deaths. Of the 598 deaths in NY State, 450 are in NYC. Observation indicated to me that travel from Wuhan and Italy to NYC ais responsible.

Africa has 4,026 confirmed covid-19 cases sprinkled throughout 47 African countries or territories. Only 97 deaths are reported but the continent is a tinderbox.

So far in 2020, we have learned that a tiny virus can bring nations to their knees. Let's observe and learn and verify to each other that more good than bad can come out of any situation when we let it. The future is bright and hopeful.

Toxin-free Smart Sugars can answer questions that drugs cannot answer. Glycoscience already impacts all healthcare but the public and most health professionals are not yet educated in the science that can provide the solution through immunology.

Mystery of the Next 2 Weeks
Apply these 3 Steps to Stay Safe and Healthy

Glycoscience NEWS Lesson #48 for 2020

03/28/2020 –

For the Next 2 Weeks:
Do not fear the evil that lurks in the darkness. As the virus spreads, the enemy will disseminate disinformation to make US to blame. War of words is evidenced in an international propaganda campaign to confuse the world in "proving" their superiority.

Nothing is as it appears and each of us views what is in front of our eyes through the glasses of our knowledge and prejudices. Critics without solutions are just critics adding to the chaos to compound fear and division. The playbook from the first garden is to blame others for their own crimes and to ensnare and use the anti-American population to echo conspiratorial stories.

In two weeks we will compare the statistics and how different countries best handle the covid-19. This will provide significant data and lessons for the future. In this evaluation, we will compare 6 countries (US, UK, Italy, Spain, Germany, and Israel) starting today (03/28/2020) and again in 2 weeks. The United States has confirmed 124,377 cases with a death count of 2,191. So far, this provides a death rate of 1.76%. Israel is at 0.33%.

Germany has a better death ratio to confirmed cases than other European countries. Let us discover why. They may have a younger average age of infection and more serious lockdown rules. Germany's coronavirus death rate is substantially lower than Italy, Spain, and the UK. Today, 57,695 Germans have tested positive with 433 deaths. That gives a death rate of 0.75%. UK's rate is 5.90%, Spain's rate is 8.17% and Italy's is 10.84%. Germany is testing some 120,000 people a week and identifying milder cases that don't end in death.

The 3 Steps to Remain Safe

Be smart, be clean, and improve your immune system. (1) Be smart – stop touching your face. (2) Be clean – soap and water on hands and surfaces you touch. (3) stop any habit that puts toxins in your lungs or stomach that lowers your immune system and consume healthful foods, including Smart Sugars.

Panic accomplishes nothing. We will survive and thrive through this virus attack and our best years are ahead.

Relax in the Virus War Zone for 2 Weeks

Day 1 – The Virus is not a Death Sentence if You Have a Good Immune System

Glycoscience NEWS Lesson #49 for 2020

03/29/2020 – HOUSTON

This is the first day of a two week count down that I call **"The Horrendous Next Two Weeks."** Dr. Anthony Fauci, director of the National Institute of Allergy and Infectious Diseases, estimated in an interview with CNN that the pandemic could cause between 100,000 and 200,000 deaths in the United States.

Panic accomplishes nothing. We will survive and thrive through this virus attack and our best years are ahead. The US has survived serious conditions in the past. The CDC reports that the flu virus kills between 12,000 and 61,000 Americans each year. During the 1918-19 flu pandemic 675,000 died.

During these next two weeks, we will learn significant lessons that will benefit US for years to come. We will gather much data. The next step is to learn how to mine the data with culture-changing knowledge.

The data that I want to study will allow us to determine what is in the blood of those affected who best resists the virus. Let us monitor the cause for flawed gene expression and develop a strategy to *Correct Flawed Gene Expression*. (Title of a training series.)

This first day of this two week period saw over 300 deaths in the US, bringing the number slightly under 2,500. Italy is nearing 11,000 with Spain under 7,000. The embers are now scattered over 48 countries of Africa with nearly 6,000 infected and more than 150 deaths.

It is easy to say, *"Relax in the middle of the storm."* This storm too shall pass and we will be stronger for it. It is so very important to be safe and ride out the storm. Do not be dismayed by those who don't. I hope I am wrong but it will be no surprise to see many suicides during the weeks ahead. Pray for your family, friends, and neighbors. Encourage and support others.

Day 2 of a 2 Week Countdown
**Plus a Lesson from "Patient Zero" in the USA
who returned to Seattle from Wuhan**

Glycoscience NEWS Lesson #50 for 2020

03/30/2020 – SEATTLE

A good sign today is that covid-19 cases are slowing in Seattle.

Patient Zero in the USA was reported January 20 – five days after he arrived at the Seattle-Tacoma International Airport from Wuhan, China. He took group transportation from the airport with other passengers. By that date, 41 people in Wuhan had been diagnosed positive but Chinese officials said the risk was low for human-to-human transmission. On 1/17, the US began checking passengers from Wuhan at airports and suspended flights on 1/23. Flights from Wuhan to Italy continued until 1/31. It was too late.

Seattle was the first US city to close offices and restaurants and instruct residents to quarantine, but it appears *"the cat was already out of the bag."* The Patient Zero story is one of confusion but we need to understand what went right and what went wrong.

At first it appeared that everything was done right. The 35-year-old man arrived 1/19 at an urgent-care clinic in a north suburb of Seattle with a slightly elevated temperature and a cough he'd developed soon after returning four days earlier from a visit with family in Wuhan. He went to a clinic, became the first positive case in the US, was taken to a biocontainment ward at Providence Regional Medical Center in Everett, Washington. His condition worsened and vacillated over the next few days. County health officials located 60+ people who come in contact with him. None developed the virus in following weeks. **By 2/21, he was considered fully recovered.** The mystery remains. Were asymptomatic people in the area? Were there carriers who were never sick?

Some researchers who traced the viral genomes of patients around the world believe someone else in the area picked it up between Jan. 15 and Jan. 19, before the traveler went to the hospital. He might have sneezed in the airport shuttle or on some surface. We have learned that covid-19 is more contagious than the flu, so wash your hands with soap and stop touching your face.

As of today (3/30/30), Washington state has had 203 deaths of the 3,170 for the nation. New York is now the epicenter with 1,338 deaths.

Today China said there is an increase of covid-19 there because travelers are bringing the virus back.

Your immune system is the wall to stop the coronavirus.

Day 3 - President Trump called these next 2 horrific weeks, "very painful 2 weeks." Then he added, *"This is going to be a very very tough 2 weeks."*

Glycoscience NEWS Lesson #51 for 2020

03/31/2020 –

I hope we are wrong about these next 2 weeks. White House coronavirus task force extended guidelines in hope of slowing the spread of the coronavirus pandemic in US. The task force predicted in the weeks ahead covid-19 could leave 100,000 to 240,000 people in the United States dead and millions infected. If other states are more like New York without measures in place to slow the spread, these totals would jump to 1.5 to 2.2 million.

If the number of deaths in China is accurate, we (that's US) just now surpassed China. This is according to an accounting published today (3/31) by Johns Hopkins University.

The CDC, WHO, and our Surgeon General have released statements that are either simply not correct or confusing and conflicting, such as, *"Americans don't need masks."* Of course, masks are needed. Why do doctors and nurses use them? But, they must be good masks and fitted properly to your nose and face. A serendipity is that the mask may keep you from touching your face so often. The mask is even harmful if: (1) it offers false security; (2) becomes a viral catcher's mitt and you touch it and then contaminate another surface – EVEN WITH GLOVES ON.

The virus spreads through droplets from coughs and sneezes directly in the air or from surfaces like cell phones, keyboards, keys, elevator buttons, shopping carts – the list goes on. The droplets may land 27 feet away unless there is a gust of wind. And, they may be active on a surface for 72 hours or longer under certain conditions. Communication makes everything happen and changing messages creates more chaos manifested in fear, confusion, and distrust. But, the experts are learning.

Covid-19 is spread by people only 3 ways – ignorance, neglect, or deliberate action. This virus may infect somewhat differently than other viruses. That fact alone, makes even the best of scientists, medical experts, and the public **IGNORANT**. Once we know what to do and don't do it's called **NEGLECT**. Let's not even consider **DELIBERATE ACTION** for now.

Johns Hopkins University reports – 858,785 covid-19 infections worldwide have already resulted in 42,000 deaths. In the US over 189,000 people have tested positive and we have had 3,900 deaths. The peak is expected here in these next two weeks.

The new frontier of medicine is Glycoscience and I wish the world could have a copy of the *Glycoscience Whitepaper* – well, you can. Ask for it.

Day 4 – 3 Ways the Coronavirus Spreads:
Ignorance – Neglect – Deliberate Action

Glycoscience NEWS Lesson #52 for 2020

04/01/2020 –

"Deliberate Action" may be the qualifying function the *"dangerously misguided"* University of Texas students took when some 70 defied good advice and went to Mexico on Spring Break. The group took a chartered flight from Austin to Cabo San Lucas a few days ago. The Texas Public Health Department confirms a cluster of 28 tested positive for covid-19 and dozens are awaiting results. The student's behavior may jeopardize many people they interacted with during the trip. Data shows that about half of the people diagnosed with coronavirus in the Austin area are between the ages of 20 and 40.

"Ignorance": The big ignorant factor is, *"You don't know where your invisible enemy is."* Doctors and healthcare professionals are overwhelmed and often do not know what to do with their patients and themselves – often overcome by the thought they could get the virus themselves. The CDC and WHO are blamed for their ignorance in not knowing what to do. Many are terrified that, because of their own ignorance, they may inadvertently pass the infection to others. Education is the CURE for ignorance as we remain vigilant.

HOT SPOTS – Florida has become a coronavirus hotspot with cases surging to about 7,000. Longtime University of Miami healthcare employee dies from coronavirus. Another man dies from coronavirus after attending an event Miami Beach event. The number testing positive has accelerated – nearly doubled in 4 days with 3,274 new cases, bringing the statewide total to 6,741 as of Tuesday evening. The state reported 857 people hospitalized and 85 deaths as of yesterday (3/31).

New York, New Jersey, Michigan, Louisiana and **Washington** are hotspots with the most deaths. Hawaii, is the 49th state to report a coronavirus death late yesterday (3/31). Wyoming is the only state without a fatality but has 120 people who tested positive.

The Ricocheting Virus – International travel has spiked cases of the coronavirus worldwide. **China** says their case increase is due to travelers bringing the virus back. **South Korea** initially flattened the curve after requiring those from some countries to quarantine – just expanded requirement to the world. **Japan** quarantined some travelers but now bars those from most of Europe. It is discussing banning travelers from the US. **China, Hong Kong, Singapore,** and **Taiwan** have shut their borders to foreigners. Europe is aflame especially in **Italy, Spain, France,** and the **UK**.

Italy's Death Toll Far Higher Than Reported
Many people are dying in Italy who were never tested for covid-19. One nursing home reported 23 people, a third of its residents, died in March. None were tested for coronavirus. Neither were 38 people who died in a nearby nursing home. Italy's official death toll from the virus today (4/1) stands at 13,155, the most of any country. But many people who die from the virus don't make it to the hospital and were never tested. A *Wall Street Journal* analysis indicates the human toll may be much greater with infections more widespread than official data indicate.

Johns Hopkins University reports – 935,817 covid-19 infections worldwide have already resulted in 47,208 deaths. In the US 215,417 people have tested positive and we have had 5,116 deaths with 2,198 in New York State. The peak is expected here in these next two weeks.

The new frontier of medicine is Glycoscience. I wish the world could have a copy of the *Glycoscience Whitepaper* that explains cell-to-cell communication – well, you can. Ask for it.

The Glycoscience Institute teaches the values of **Hippocrates, the Father of Medicine**. His value system started with, **FIRST DO NO HARM** and **LET FOOD BE THY MEDICINE AND MEDICINE BE THY FOOD**.

Day 5 – The Virus is a Killer – But FEAR NOT – It is unbelievably WEAK

A look at the virus' Vulnerability – Multiple ways to destroy it!

Glycoscience NEWS Lesson #53 for 2020

04/02/2020 –

Today's Lesson is to understand the vulnerability of the virus – not just coronavirus but all viruses. Pick your weapon and let us destroy the virus that has invaded your territory.

One of the invisible enemy's weapons is that it is INVISIBLE – it can sneak-up on you. The only power the virus has is that which we give it through IGNORANCE – NEGLECT – or REBELLION. Rebellion can take the form of terrorism or the act of simply ignoring sound advice. Neglect is not exercising wisdom of the knowledge you have. Ignorance is the lack of knowledge. People perish because of lack of knowledge.

Because it is a new mutation, scientists have been remarkably ignorant about how it functions. We have learned during the last few weeks that the tiny coronavirus can bring the world to its knees. It is a silent killer. So, we cannot see it – it sneaks-up on us -- and can strong-arm nations by killing at least thousands of people (to-date) and shutting down countries. Many who have ignored the virus have died. We better pay attention, but fear not. This pandemic too will pass.

Let us look at the weaker side of the virus to stop the spread and defeat this wicked enemy. People are probably washing their hands more than any time in history. That is good when you use soap because the surfactant dissolves the virus. It is even more easily dissolved when you use hot water because viruses dissolve at 133 degrees Fahrenheit (56 Centigrade).

I enjoy investing (instead of spending or wasting) a few minutes in a hot sauna where I cautiously and slowly breathe deeply. The temperature of 140°F can be beneficial in destroying the virus on and in your body. A steam room with the temperature 140°F can be as beneficial when you cautiously and slowly breathe deeply. Be safe if you choose to attempt to achieve the same results with a make-shift effort such as sticking your nose near a teakettle spout or over boiling water.

Certain essential oils melt the virus like ice cream in the sun. We have used an atomizer in the rooms that may be very effective. Use sanitizer cloths to wipe down keyboards, keys, credit cards, cell phones, touch pads, flat surfaces, buttons on public elevators, door handles, steering wheels, and you can add to the list. A UV light wand may be another tool to melt the virus. Take precautions – don't look at the light. And, do not use the UV light on certain plastics. It can melt and produce plasticizers which are carcinogenic.

TODAY'S WORLD REPORT for 04/02/2020: Today marked the day when more than 1 million people were confirmed to be infected with covid-19 and more than 53,000 confirmed deaths to-date; US confirmed cases is more than 245,000 with 6,000 deaths. Europe has lost tens of thousands of people. 52 countries in Africa have 7,200 cases with 275 confirmed deaths.

You will see more good than bad come out of this pandemic.

Day 6 – Nothing is as it Appears – Sorting out Right from Wrong, Dumb from Smart, and Good from Bad

Glycoscience NEWS Lesson #54 for 2020

04/03/2020 –

The tiny coronavirus has brought planet earth to its knees with a pandemic that will teach us, help us correct some things, and greatly benefit humans for generations to come. But there is still much chaos, misinformation, and disagreement based in ignorance. Experts agreed today that covid-19 models are rather worthless.

This pandemic is a giant scientific study – My personal definition of science is: the observation and experimentation of the structure and action of any part of what is.

I still believe scientists and health professionals are looking through the wrong end of the microscope. Let all the experts, let all the local, state, and national experts arrive at their conclusions – compare all their conclusions based on RESULTS. We are in a worldwide petri dish to reveal our character which will change our culture. How we respond instead of react – how we resolve problems instead of criticize – reveal the character of the individual, community, or country. The culture of our country is about to change. My concern is that there will remain two polarizing cultures – with some calling good, "bad," smart, "dumb," and right, "wrong." The results of this petri dish culture test will be self-evident but many will refuse to acknowledge what it is.

Remarks about Reader's Requests: Thanks for the comments and questions concerning Lesson #53 yesterday about the multiple ways to destroy viruses. The questions were based around " (1) What sugars, (2) what essential oils, and (3) what UV lights do I recommend." While I am an authority on glycans (Smart Sugars), I do not consider myself an expert on essential oils (though my wife is an authority but she does not sell them) nor am I an authority on UV lights but fairly knowledgeable.

(1) WHAT SUGARS? As I have outlined in several books and articles, there are some 27 specific sugars found in nature that are remarkably beneficial to good health. I recommend a complex of these sugars and mannose (these sugars are not sweet). And for the sweet-tooth, I recommend (especially for diabetics) replacing regular sugar (which lowers the immune system) with Trehalose or a 5 times sweeter than regular table sugar product that is diabetic-friendly with significant health benefits. I formulated this sweetener called Sweet and Healthy

X5. Another complex I formulated is T/C+ which may be beneficial for the advanced diabetic. T/C+ also offers significant neurological support for those with Alzheimer's, Parkinson's, Huntington's, MS, and ALS. We had a research paper published through Ben-Gurion University in Israel that the functional sugar in X5 is an antidepressant – for such a time as this.

(2) WHAT ESSENTIAL OILS? Essential oils should be certified organic or "wild crafted" – free from all pesticides. Some essential oils are reported to destroy viruses better than others. From what little I know about essential oils – it appears to me that the top 7 may be thyme, lemon, peppermint, lavender, eucalyptus, pine, and rosemary.

(3) WHAT UV LIGHTS? UV lights or wands: The Portable UV Sterilizer I have is reported to kill bacteria and viruses. The "test results" are impressive but I do not have verification evidence.

TODAY'S WORLD REPORT for 04/03/2020: Covid-19 confirmed 1,099,389 infected and 58,901 confirmed deaths; US cases is 277,953 with 7,152 US deaths. Europe has lost tens of thousands of people. 52 African countries report 8,119 cases and 333 confirmed deaths. There are many hot spots and tinderboxes. Pray for wisdom to best address this crisis.

The new frontier of medicine is Glycoscience. You can help tell the world about Glycoscience by giving others a copy of the *Glycoscience Whitepaper* which explains cell-to-cell communication.

Day 7 – This Next Week will be Toughest Week Yet

Glycoscience NEWS Lesson #55 for 2020

04/04/2020 –

One week ago (3/28), I announced the *"Mystery of the Next 2 Weeks."* The the next day, I called it, *"The Horrendous Next 2 Weeks."* The following day, President Trump announced, *"This is going to be a very very tough 2 weeks."* Today, he said, *"There will be death"* and then added that this coming week will be the *"toughest week. ... We will move heaven and earth to safeguard our great American citizens."*

One Week Comparison

Here is a quick comparison of where the world and the US were at the beginning of the 2 weeks and today. The 1st day of this 2 week period saw over 300 deaths in the US bringing the number slightly under 2,500. **Today, the 7th day, saw 1,354 deaths in the, bringing the number to 8,496.** On the 1st day, Italy was nearing 11,000 with Spain under 7,000. **Today, Italy's official count is 15,362 and Spain is 11,947.** One week ago, the infectious embers were scattered over 48 countries of Africa with nearly 6,000 confirmed infected and more than 150 deaths. **Today there are 52 African countries confirming 8,814 infections with 383 deaths.** This count is remarkable and wonderfully lower than expected. Yet, many potential hot spots on the planet remain tinderboxes.

Just ahead

This coming week will provide new understanding of why the most vulnerable are the most vulnerable. We will examine why the native American is vulnerable to the virus and what we can do about it. We will examine why the diabetic is more vulnerable and what we can do about it.

The new frontier of medicine is Glycoscience and I wish the world to have a copy of the *Glycoscience Whitepaper* – well, you can. Ask for it.

Day 8 – What have we Learned? What will we Learn this Week?

Glycoscience NEWS Lesson #56 for 2020
04/05/2020 –

Garnered information from 182 nations on how they deal with covid-19 will impact a cultural change in the medical community and help save lives for generations to come. Comparing RESULTS will provide much needed data.

The population of 2 countries are similar. One has 5,687 confirmed cases with 35 deaths. The other country has 363 confirmed cases with 5 deaths. What did they do differently?

Australia and Taiwan are both islands of about 24 million with strict controls over who comes into the country. Both countries have strong trade and transportation links with China. How has Taiwan been able to better keep the virus under control?

Conoravirus is covid-19 and is also SARS-CoV-2. Scientists have trouble agreeing on what to call things. In 2003, Taiwan was among the hardest hit areas with the SARS (severe acute respiratory syndrome) outbreak. More than 150,000 people were quarantined on Taiwan and 181 people died. Taiwan learned how to respond – how to react faster and to take the danger more seriously. Taiwan quickly put in place over 120 action items to protect the public health. Border control was not the only option.

Taiwan's leadership was taking action while congresses and committees in other countries were still debating to take action or not to take action. Johns Hopkins University reported a January study that Taiwan was one of the most at-risk areas. But they did not build into the equation Taiwan's actions. The early decisive measures were to ban travel from parts of China, stop cruise ships from docking at its ports, and establish strictly enforced home quarantine orders while avoiding strict lockdowns that characterized the dictatorial response of other countries. They ramped up face-mask production to ensure local supply, rolled out testing and retesting for covid-19 and announced punishments for spreading disinformation about the virus.

A major key to success against covid-19 was the country's rapid actions. Its experienced teams were quick to recognize the crisis and implement emergency management structures they drew from their 2003 textbook.

Last week Taiwan lifted their ban on exporting face masks that had ensured its domestic supply and offered to donate 10 million masks to the United States, Italy, Spain and 9 other countries with whom they have diplomatic relations. It is noted that Taiwan is not a member of the WHO because China blocks or dictates its participation in international organizations.

The new frontier of medicine is Glycoscience and I want you to have a copy of the *Glycoscience Whitepaper* – well, you can.

Applied Glycoscience is the toxic-free answer to questions that drugs cannot answer. Glycoscience already impacts all healthcare but the public and most health professionals are not yet educated in the science that can provide the solution through immunology.

TODAY'S WORLD REPORT 04/05/2020: Covid-19 confirmed 1,274,923 infected worldwide with 69,487 reported deaths. US confirmed 337,635 infected with 9,647 deaths - 54 African countries with 9,297 confirmed infected - 446 reported deaths of which 113 deaths reported last 24 hours.

Day 9 – How a Placebo Can Save the Day; Exciting Possibilities Against ALL Viruses

Glycoscience NEWS Lesson #57 for 2020

04/06/2020 –

Hydroxychloroquine, the anti-malaria drug, is thought to be a possible wonder-drug against covid-19. President Trump says that at least we need to try something that might work. The medical battle is that Hydroxychloroquine has not been "tested" in double-blind placebo trials for the new virus. It is, however, one of the leading causes of drug overdose in malaria-prone countries and a lupus or rheumatoid arthritis medication that can lead to lethal consequences.

I have ethical and moral concerns about using Double-Blind Placebo trials. You have a group of people about to die and you think there is no cure. You give half of them the drug that might save their lives and you withhold the drug from the other half of the trial group and give them a placebo that you "know" won't work and you let them die. This archaic medical practice is the standard medical procedure imposed by the FDA. This is one of the reasons the FDA must be updated. Is it possible to "Do no harm" while playing God with Double-Blind Placebo trials? Hydroxychoroquine qualifies for another archaic FDA rule – the LD50 level (Lethal Dose 50). A tiny amount of hydroxychoroquine will easily kill 50% of the animals in a study.

The chaos and disagreements within the medical powers surrounding the President are centered around Double-Blind Placebo trials. The placebo is usually a sugar or something the medical establishment "knows" is ineffective. They are considering vitamin C as the placebo. I think it is a wonderful and great plan to immediately run Double-Blind Placebo trials on those at death's door with covid-19 – just let me formulate the placebo. My formulation would consist of natural vitamin C (NOT ascorbic acid) and let me choose the Smart Sugars.

COVID-19 DOUBLE-BLIND PLACEBO TRIAL
These Double-Blind Placebo trials could be conducted immediately. They could be conducted with half of any size group or multiple groups receiving the hydroxychoroquine and the other half receiving my Applied Glycoscience Sugar. The moral issue is gone because, with the research we have already conducted, I am convinced that the non-toxic Placebo Group will have strikingly better results than the highly toxic drug or combination of drugs. I have reviewed science papers where the solution used in the studies failed – not because it did not work but because those who designed the test used a smaller amount so that it would fail. FOR THE COVID-19 DOUBLE-BLIND

PLACEBO TRIAL with my formulation as the placebo, I would determine the amount of "placebo" used. Hint, it is difficult to use too much because it is TOXIN-FREE.

Senator Ted Cruz has written a Bill to restructure the FDA – his Bill is in my book, *To Kill A Rat*. Senator Cruz is on the record, *"We need to tear down the barriers blocking a new era of medical innovation, and the primary inhibitor is the government itself. It's past time to unleash a supply-side medical revolution, so that instead of simply caring for people with debilitating diseases, we cure them ... and embrace a culture of innovation."*

The coronavirus has provided the opportunity to discover WHAT REALLY WORKS!

The new frontier of medicine is Glycoscience and I want you to have a copy of the *Glycoscience Whitepaper* – well, you can.

The Glycoscience Institute teaches the values of **Hippocrates, the Father of Medicine**. His value system started with, **FIRST DO NO HARM** and **LET FOOD BE THY MEDICINE AND MEDICINE BE THY FOOD**. The Hippocratic Oath is no longer widely practiced and it is time to get back to it.

Applied Glycoscience is the toxic-free answer to questions that drugs cannot answer. Glycoscience already impacts all healthcare but the public and most health professionals are not yet educated in the science that can provide the solution through immunology.

TODAY'S WORLD REPORT 04/06/2020: Covid-19 confirmed 1,347,803 infected worldwide with 74,807 reported deaths. US confirmed 368,196 infected with 10,863 deaths - 55 African countries with 10,253 confirmed infected - 490 reported deaths.

Day 10 – What is that Light at the end of the Tunnel? – Pray it's Not a Train

Let us use an integrative approach to come along side our great doctors and healthcare professionals and help them be more efficient and effective.

Glycoscience NEWS Lesson #58 for 2020

04/07/2020 –

The mystery of that Light flickering at the end of the tunnel may be different than any of us thought. Whatever the reason, it will most likely change the medical culture. Light is the best disinfectant – it reveals and changes corruption, infection, viruses, characters, and cultures. For these problems, it is a train.

The worldwide petri dish of the coronavirus is on display for all to see. How each territory, each region, state, country, and continent responds to the virus is a medical and social lesson on how to best deal with future infections for generations to come. The culture, the character, of each is being revealed. The bell curve for each region is to be examined by scientists and the general public. Each will view the results through the glasses of education and prejudicial tendencies.

When I was a child, I envisioned the most powerful position in the world was the doctor because it is the doctor who tells even the President and all the leaders what to do. Today, the patient has taken that authority from the doctor. It is the patient who is now calling the shots (pun intended).

The medical mind-set is to develop a new wonder-drug for each of more than 5,000 viruses and more than 300,000 known pathogens. It is the growing mind-set of the patient that says, "*There's got to be a better way.*" There are apparently dozens of ways to destroy viruses outside of toxic drugs – but the medical community in general chooses to ignore what works unless it is to cut, burn, or use a toxic drug.

An example is that today Boris Johnson, UK PM, is in intensive care in England with covid-19. Reports are that more than a million people died from scurvy between the time they knew what to do and the time they actually did it. Lemons were so effective for physical health and military battle that the British sailor became the "*limey.*" The FDA, yet, does not allow us to put a label on the lemon that it treats and CURES scurvy.

Still today, the British apparently do not treat their PM with something natural that has worked for centuries. Okay, I know covid-19 and scurvy are different. And, I am aware of the many attacks and fake news on the efficacy of lemons and limes. But, there appears to be solid research published in Virology Volume 485 November 2015 pages 199-204 showing that lemon citrate dissolves the norovirus particles and renders the virus disabled. The fact is that most of us probably have a deficiency of vitamin C – so adding a lemon or two each day to our diet may be a wise move. This is a "Do no harm" act for better health.

Light is shining on the WHO and WHO is damaged for its deception and flawed reporting on the coronavirus. Let us not throw out anything that works to benefit human health. In this pandemic, the patient is shining a bright light on what works. Keep those lights shining and let me know what works best in addition to supporting the immune system.

New frontier of medicine is Glycoscience, see *Glycoscience Whitepaper*

The Glycoscience Institute teaches the values of **Hippocrates, the Father of Medicine**. His value system started with, **FIRST DO NO HARM** and **LET FOOD BE THY MEDICINE AND MEDICINE BE THY FOOD**. The Hippocratic Oath is no longer widely practiced and it is time to get back to it.

Applied Glycoscience is the toxic-free answer to questions that drugs cannot answer. Glycoscience already impacts all healthcare but the public and most health professionals are not yet educated in the science that can provide the solution through immunology.

TODAY'S WORLD REPORT 04/07/2020: Covid-19 conformed 1,430,141 infected worldwide with 82,119 reported deaths. US confirmed 398,809 infected with 12,895 deaths - 56 African countries 10,950 confirmed infected with 535 reported deaths.

Millions of Lives May be Spared in Africa in a Very Strange Way

Glycoscience NEWS Lesson #59 for 2020

04/08/2020 –

Indication is that Africa is a tinderbox awaiting detonation. Fifty-six African nations are infected with coronavirus. The small number of confirmed embers are scattered to the four corners of Africa – that is enough to kill millions of people. The embers have not ignited as expected.

Fragility of weak health systems and lack of sanitary conditions make for dry tinderbox conditions ready for outbreaks that could destroy the continent. The other day, the African Union Bureau and AU Commission held a teleconference with African leaders. They received a presentation from WHO Director-General Dr Tedros Adhanom Ghebreyesus and the Africa Centres for Disease Control and Prevention (CDC) Director Dr John Nkengasong, who provided an update on the state of the pandemic in Africa and across the world. The WHO leadership has previously coordinated Chinese communist influence into African nations. Should WHO be as influential with African leaders as its leaders attempted to be with US – millions will die in Africa as the result and China will control the continent. Director-General Dr Tedros has praised China for its "transparency" on the virus.

It may be too early to know for sure if this is what is actually happening.

It appears that a delightful surprising twist may be taking shape to protect our brothers in Africa. What countries are hardest hit with malaria? Several of the African countries including Guinea, Botswana, Burundi, Zambia, and Malawi are reportedly affected with malaria.

Hydroxychloroquine is a poison drug and in my opinion is NOT the best way to treat coronavirus; however it appears to be the "standard of care" for treating malaria around the world. Is it a preventative drug for coronavirus?

I studied my charts of the 56 African countries where covid-19 has landed and I looked specifically at Guinea, Botswana,

Burundi, Zambia, and Malawi. Here is what I learned: As of today (4/8/2020)

	Confirmed cases	Reported deaths
Guinea	144	0
Botswana	6	1
Burundi	3	0
Zambia	39	1
Malawi	8	1

New frontier of medicine is Glycoscience,
see *Glycoscience Whitepaper*

TODAY'S WORLD REPORT 04/08/2020: Covid-19 conformed 1,484,811 infected worldwide with 88,538 reported deaths. US confirmed 432,132 infected with 14,817 deaths - 56 African countries 11,662 confirmed infected - 576 reported deaths.

What we learn from the Invisible Killer will help prepare generations to come. It's biggest lesson is *"An ounce of prevention is worth a pound of cure."*

Tragedy in New York – Investigation in Africa

Glycoscience NEWS Lesson #60 for 2020

04/09/2020 – HOUSTON

It's not normal

Lockdown, self-quarantine, no travel, social distancing, and the power has been off for hours – just now turned back on after a storm in Houston.

The world has been brought to its knees by a tiny invisible entanglement of protein. Chaos, confusion, lies, and deception are in the petri dish to study our culture and character. The bright Light is needed to make uncommon sense out of this darkness. Manipulation of unparalleled proportions is in clear sight. Contamination in the petri dish is growing and being revealed. It is up to each one of us to take authority to clean our territory. An awakening, a cleansing is coming – and this is a good thing! What was meant for destruction will destroy those attempting to destroy. There will be loud screams because things are not working out as many had planned.

The Malaria Connection

An investigation has begun to determine if many in Africa who were previously treated for malaria with hydroxychoroquine were (accidently) immunized against the coronavirus.

Hydroxychloroquine is an antil drug. It is derived from quinine and cinchonine, which was originally taken from the bark of the cinchona tree from Peru. The bark from this tree is reported to modulate the immune system. Quinine is in the class known as antimalarials. I remember from my very young days knowing what quinine was and the term still resinates in my mind, "*bitter as quinine.*" Another term was, "*A little bit of sugar makes the medicine go down.*"

My charts of the 56 African countries has not changed much the last 24 hours. 60 deaths in several African countries but none reported from covid-19 in Guinea, Botswana, Burundi, Zambia, and Malawi.

Confirmed cases		Reported deaths
Guinea	194	0
Botswana	13	1
Burundi	3	0
Zambia	39	1
Malawi	8	1

Tragedy in NY – over 7,000 deaths

New York City, New York State, and New York people were gripped with growing despair. Previous underlying conditions have left them vulnerable to chaos stoked by worry, doubt, fear, anxiety, and frustration. And, that was before cronoavirus. A people ravaged by the virus have compounding challenges when underlying conditions manifest as diabetes, high blood pressure, and a poor immune system.

As Stupid as it gets –

A young lady was arrested because she was walking too close to her boy friend – she was put in a jail cell with 90 other women.

TODAY'S WORLD REPORT 04/09/2020: Covid-19 confirmed 1,602,885 infected worldwide with 95,745 reported deaths. US confirmed 466,299 infected with 16,686 deaths - 56 African countries 12,721 confirmed infected with 636 reported deaths.

Corona is a Crown of Thorns – It is a Dark Day – 102,000+ Dead so Far

Glycoscience NEWS Lesson #61 for 2020

04/10/2020 – HOUSTON

The coronavirus is literally a Crown of Thorns. Corona is the acclaimed king of sickness, death, and economic devastation to earth. It has brought the world to its knees for such a time as this. Its name comes from the protein protrusions on its surface like a crown of thorns. The thorns become snared onto the surface of weak living cells.

The Crown of Thorn virus is the greatest example of evil personified – worse than a parasite which only feeds on and lives off of another life – this virus has a single helix RNA (genetic material) wrapped in entangled protein. It literally rapes a living cell. It penetrates an appendage into the cell and injects its RNA. It now owns the cell that will never reproduce itself. The cell will now only reproduce a virus. The new virus too is powerless on its own and only has the power it is given. The virus is stopped when the immune system tells it to stop.

Lack of integrity of the immune system gives way to a dark day. We can forever learn without application of wisdom. The world is focused on the problem instead of the solution. We can keep looking through the wrong end of the microscope to seek scientific explanation of reality of what appears to be four sub-groupings (there may be more) of the coronavirus – alpha, beta, gamma, and delta. But that is for a later examination and discussion – if at all. So, the problem is NOT THE VIRUS! The problem is lack of integrity of the immune system – lack of integrity of cell signals. Communication makes everything happen. The world is looking in the wrong place and getting more darkness.

Why did the Crown of Thorns come today? You ask, "*What is the significance of the crown of thorns*?" Today is the day of remembrance of what happened more than 2,000 years ago, when a painful crown of thorns was thrust upon the head of Jesus in a final mockery just before His crucifixion. Today is a dark day in human history when at 3 o'clock in the afternoon on Passover, the 14th Day of Nissan, Jesus - (Yeshua Hamashiach) The King of Kings and Lord of Lords was crucified on the stake at the very moment the Passover Lamb had been sacrificed since Moses brought them out of Egypt, Africa. Africa has significance and will bless the world in ways that may soon be revealed.

My charts of the African countries have not changed much the last 24 hours. 35 reported new coronavirus-related deaths in all African countries, but only one reported death in the malaria troubled countries of Guinea, Botswana, Burundi, Zambia, and Malawi where quinine seems to prevent the Crown of Thorns virus from killing their people.

As Stupid as it gets –
Police arrested people in a church parking lot with a $500 fine for attending church where they never got out of their cars. But its okay to go shopping in the same area.

TODAY'S WORLD REPORT 04/10/2020: Covid-19 confirmed 1,698,416 infected worldwide with 102,764 reported deaths. US confirmed 501,419 infected with 18,769 deaths.

Between Friday and Sunday

Waiting – Day of Darkness – US Passes Italy in Deaths – Africa Spared?

Glycoscience NEWS Lesson #62 for 2020

04/11/2020 – HOUSTON

A lot is going on in the darkness! Waiting is so difficult. We will examine, in a moment, how an amazing strength can come from waiting.

Today US passed Italy in coronavirus-related fatalities. (Italy has about 1/5 the US population) Still looking through the wrong end of the microscope, scientists are "confident" that a vaccine for the coronavirus could be ready someday. New York's governor and New York City's mayor are feuding. Queen Elizabeth's first-ever Easter message: *"Coronavirus will not overcome us. ... As dark as death can be, light and life are greater. May the living flame of the Easter hope be a steady guide as we face the future."* The queen's title is, *"Defender of the Faith and Supreme Governor of the Church of England."* New Orleans' French Quarter sits empty. Rats swarm the streets in search for food. The USS Roosevelt has 550 confirmed covid-19 cases. Cruise ships are out of business for now. The virus is more infectious than we thought. Severe storm warnings with coming tornadoes. Chernobyl wildfires reignite radiation fears. Now is the time of greatest risk. Pray the death angel will Passover and that the worst is past and that we overcome.

Caution with chloroquine, hydroxychloroquine, or quinine

All 3 are toxic. Some countries in Africa that have used a quinine -based drug for malaria seem to have a high resistance to the coronavirus and, therefore, have a much lower death rate than other countries. During World War I, quinine played an important role in keeping our troops safe from malaria. Quinine comes from cinchona bark. Drug companies first synthesized quinine in 1944 but found it more expensive to synthesize. It is a bitter alkaloid in a class with nicotine. Because it may be a cheap pathway to treat coronavirus and because the President has suggested its use, expect a serious campaign against its use.

Strength from waiting

There is a wonderful old Scripture that goes something like this: *"They who **wait** upon the Lord shall renew their strength; they shall mount up with wings as eagles; they shall run, and not be weary; and they shall walk, and not faint."* The English language does not have a word for this Hebrew word "wait." The closest we have is "waiter." This is an active wait, not a passive wait. This word "wait" has a deep meaning inviting the weaker to entwine around the stronger to receive the strength of the stronger – as a vine wraps itself around a huge oak tree and is not only able to withstand the storm but, after the storm, finds itself wrapped even tighter. WAIT – the Light is coming!

Africa

My charts of the African countries have not changed much. 45 reported new coronavirus-related deaths in all African countries today (4/11) with only one reported death in the malaria-troubled countries of Guinea, Botswana, Burundi, Zambia, and Malawi where quinine seems to prevent the Crown of Thorns virus from killing their people.

You can't make this up –

I mentioned police arresting church-goers in their cars in the parking lot for not staying home while shopping is okay. And, about the lady arrested for not distancing herself while she and her boy friend went for a walk. They put her in a jail cell with 90 others. Meanwhile, violent criminals were released to make room in the jails. Our freedom must not be stolen by a virus or thieves.

The Glycoscience Institute teaches the values of **Hippocrates, the Father of Medicine**. His value system started with, **FIRST DO NO HARM** and **LET FOOD BE THY MEDICINE AND MEDICINE BE THY FOOD**. The Hippocratic Oath is no longer widely practiced and it is time to get back to it.

TODAY'S WORLD REPORT 04/11/2020: Covid-19 confirmed 1,777,517 infected worldwide with 108,867 reported deaths. US confirmed 529,887 infected with 20,604 deaths.

Necessity is the Mother of Invention – Let's Prioritize Necessities to Rid the World of the Crown of Thorns

Glycoscience NEWS Lesson #63 for 2020

04/12/2020 – HOUSTON

In the weeks ahead, necessity will drive innovations to solve the pandemic. Some of these solutions will be "out of the box" and resisted by the elite but welcomed and used by much of the public. Remember, the patient is calling more of the shots. Let us prioritize the necessities and we will learn several innovations.

This Easter finds us in a petri dish that will soon expose the infections, the corruption, in the bureaucracy of science and government. The great medical minds are looking through the wrong end of the microscope.

When I was a 10-year-old boy, I enjoyed listening to the national *Wize Kid* radio program. A boy my age was asked, "*What is the most important word in the English language?*" The kid responded, "*Redemption.*" I immediately agreed and pondered that *redemption* was God's greatest innovation since creation.

The world needs redemption from this pandemic. There are 10 words from over 2,000 years ago that I wish to apply in answer to where medical scientists are looking – "*Why are you looking for the living among the dead?*" The world is focused on death instead of life as their eyes are fastened on the devastation. The image the world sees is that Goliath is loose and killing people. Instead, let us fasten our eyes on the solution – various possibilities because in reality, the virus is weak, fragile, and easily destroyed.

The coronavirus is spread ONLY through IGNORANCE, NEGLECT, or INTENTION.

Africa

My charts of the African countries have not changed the last 24 hours. 18 reported new coronavirus-related deaths in all African countries today (4/12) with 0 reported deaths in the malaria-troubled countries of Guinea, Botswana, Burundi, Zambia, and Malawi where quinine seems to prevent the Crown of Thorns virus from killing their people.

Learn about the NEW FRONTIER OF MEDICINE – Glycoscience, see **Glycoscience Whitepaper**.

TODAY'S WORLD REPORT 04/12/2020: Covid-19 confirmed 1,850,527 infected worldwide with 114,245 reported deaths. US confirmed 557,571 infected with 22,108 deaths.

10 Ways to Destroy any Virus

Seek professional help when melting a virus – Do not follow these instructions unless you take full responsibility for your own health

Glycoscience NEWS Lesson #64 for 2020

04/13/2020 –

Does anyone really know what is going on? The world is turned shutdown by an invisible enemy. The enemy contains a fragment of genetic RNA coding. But, it is much more than that. The virus may be the best example of a tangled misfolding of protein. If proper folding of the proteins could be achieved, it could possibly address all neurological challenges and impact resistance to all viruses.

Scientists are at odds if a virus is alive or dead. A virus resembles a cell but cannot eat nor replicate on its own. It normally has a one-legged gene (single helix RNA). It does not have cell structure nor its own metabolism. It requires a host cell. The virus has a fragment(s) of genetic material wadded into an entanglement of protein to function as the messenger of a packet of bad information awaiting transfer to a weak living cell.

Cell signal integrity – proper communication between cells makes all good things happen in the body. The lack of cell signal integrity (miscommunication) is the cause for the misfolding of proteins. It is the misfolding of proteins that is the cause of or accentuation of all neurological diseases, including Alzheimer's, Parkinson's, Huntington's, ALS, and MS. Certain sugars in the body assist in the folding of the proteins and, when these sugars are added to the body, they support neurological functions for mental and motor skills.

The genetic coding in the virus contains a splice of incomplete information, with a package of instructions. A single strand of DNA can compact an unbelievable amount of data or corrupted data. Let me attempt to briefly explain. When the sperm fertilizes the egg, those two cells become one stem cell that contains not only the complete blueprint for development of the new life but the data of both ancestors back to the beginning. And, the RNA verifies that we all descended from one woman. That is science! Here is an example of a message that makes for opposite results: "Do not go." and the integrity is lost to "Go." Now, put a wad of corrupt data fragments in an entanglement of protein and you have one mean virus.

Seek professional help when melting a virus – Do not follow these instruction unless you take full responsibility for your own health

It does not take much to cause the misfolding of protein. Like getting heat close to a sheet of plastic – it quickly crinkles to look like crape – that is crape as in crape paper. That is how fragile is the virus. We need to let the world understand you can melt the virus on the spot even in your body. We sanitize surfaces and wash our hands. Soap and hot water melts the little enemy. A hot sauna or hot steam room may be an enjoyable way to melt the virus. It melts at 133°F. Nasal flush with weakened hydrogen peroxide solution may clear sinuses and melt the virus. Zinc is antiviral. UV light on surfaces and everything you touch can melt the viruses. Copper may be antiviral. Cinnamon is reported to be antiviral. Hey, that may be why we are having such remarkable response to our Pilot Studies using T/C+ with Parkinson's, Alzheimer's and other neurological challenges. I formulated T/C+ with research indicating it helps properly fold the proteins in the body.

Something strange is going on, especially in Africa, with infections (or lack of infections) of covid-19. We observed that the Africans where malaria is a problem and have used quinine – the people seem to resist the coronavirus. The US military used quinine against malaria and introduced its benefit to Africa decades ago.

My charts of the African countries have not changed much in the last 24 hours. All of Africa has recorded a total of 840 covid-19 related deaths as of today (4/13) including a total of only 6 deaths in the malaria-troubled countries of Guinea, Botswana, Burundi, Zambia, and Malawi where quinine seems to prevent the covid-19 from killing their people.

TODAY'S WORLD REPORT 04/13/2020: Covid-19 confirmed 1,920,985 infected worldwide with 119,687 reported deaths. US confirmed 582,580 infected with 23,628 deaths.

10 More Ways to Destroy the Virus

While we are eradicating the virus – let's lower your risk!

Glycoscience NEWS Lesson #65 for 2020

04/14/2020 – HOUSTON –

The world is still in a daze at how quickly things can change. Let's pull out all the stops and win the battle. High-tech and low-tech are at play and both can eradicate the virus.

I was born on a Missouri farm and had no running water or electricity. We kept food cool by lowering it in a bucket into a cistern where it was cooler. My father told me that if we put a silver dollar in the milk jar, the milk won't spoil as quickly. That was when I learned that silver would destroy bad stuff. After we got electricity, I would sit on the floor close to the radio and enjoy episodes of the Lone Ranger. I associated his silver bullet to abolishing evil. I was intrigued by the fact that silver could protect. I had proof that it was a solid science. But, today "experts" scoff. Silver is antiviral and antibacterial. Early this morning, I drink a "swig" of colloidal silver on an empty stomach. Tonight before I go to sleep, I will drink a "swig" of colloidal silver on an empty stomach. A dear friend made sure that I was protected from the covid-19 by colloidal silver and shipped to me a case from Canada. Thank you!

High-tech

A news flash just announced the invention of a quick, safe, advanced method for covid-19 testing. Scientists in Haifa, Israel also announced that placenta-based cell-therapy used on the first 6 critically ill covid-19 patients had a 100% survival rate. Great! Stem cell treatment will be used tomorrow to accomplish medical wonders. At the Glycoscience Institute, we teach how to proliferated stem cell production in your own body with glycan technology. Doctors and others may request training.

Low-tech to Lower Your Risk

High-tech is great but low-tech is simple, immediate, and cheap. Correcting habits may be the most beneficial thing you can do to overcome the virus pandemic. Correcting bad habits and replacing them with good habits may even put money in your pocket. Covid-19 can destroy your lungs – so it makes a lot of sense to stop smoking and start detoxing your lungs. Regular table sugar lowers your immune system and you don't have time for a weak immune system – in fact, you may have no time at all. So, change your bad habits that makes for higher risk and lower risk by drinking more clean water to flush out the toxins.

Lowering your risk can be enjoyable. I have learned to replace soft drinks (they lower your immune system) with a delightful drink of freshly squeezed lemon juice into 8 to 16 ounces of clean water sweetened with what may be the most healthful natural sugar/sweetener on the planet. **Best lemonade you ever tasted.** You can learn more about it at www.TryX5.com.

Antiviral compounds are in lemons and they are known to help fight the flu with natural vitamin C. Citrus fruits contain antiviral agents. In addition to lemons and limes, other foods have antiviral benefits including walnuts, pomegranates, onions, tomatoes, certain mushrooms, coconuts, different teas. Bright colorful fresh fruits and vegetables are normally rich in nutrition with high-antioxidant activity.

Seek professional help when melting a virus – Do not follow these instruction unless you take full responsibility for your own health.

In the midst of the carnage, something wonderful is going on around the world, especially in Africa, with halting infections of covid-19. We observed that the Africans where malaria is a problem have used quinine – now the people seem to resist the virus. The US military introduced the benefits of quinine to Africa decades ago.

My charts of the African countries have not changed much in the last 24 hours. All of Africa has recorded a total of 897 covid-19-related deaths as of today (4/14), including a total of only 6 deaths in the malaria troubled countries of Guinea, Botswana, Burundi, Zambia, and Malawi where quinine seems to prevent the covid-19 from killing their people. No recent deaths in those 5 countries.

TODAY'S WORLD REPORT 04/14/2020: Covid-19 confirmed 1,981,239 infected worldwide with 126,681 reported deaths. US confirmed 609,240 infected with 26,033 deaths.

In the midst of the carnage, something wonderful is going on around the world, especially in Africa.

Desperate Search for Common Sense

Interesting FACTS just came in that will Change Everything!

Glycoscience NEWS Lesson #66 for 2020

04/15/2020 – HOUSTON

The petri dish in which our culture is thrust reveals the mindless approach some "leaders" are forcing upon the people. The public responds and makes known that the cure cannot be worse than the disease. The virus is the second biggest threat to humans. Humans are the number one threat to humans.

I am so excited to learn the benefits that will come out of all this chaos and confusion. Some areas are yet to see the worst. How individuals, states, and nations respond tell us much about the charter and culture of those responding. Neighbors are helping neighbors – selfless first responders are putting their lives on the line to help others. They are at great risk. Their focus is often on helping others at their own peril and not taking proper precautions for themselves. You can help lower that risk by passing on sound advice.

Our first wave of covid-19 came through IGNORANCE in not knowing what to do. The second wave may come through NEGLECT. Yesterday (4/14), was the largest death toll for a day after reaching, what appeared to be, the top of the bell curve. The total confirmed deaths for one day in the nation was 2,405, according to case tracking by Johns Hopkins University.

Observing the nations and areas where the least number of covid-19 cases manifested provides very interesting data. Here are the statistics frozen in a moment of time today: US confirmed cases 639,055 with 30,925 deaths; Italy with ~1/5 of the population of the US has 165,155 confirmed cases and 30,844 deaths; Israel with 12,501 cases and 130 deaths; Taiwan confirmed a total of 395 cases and 6 deaths. Note: Taiwan learned from the previous infestation of SARS in 2003 and the WHO (because of its support of China), did not assist Taiwan.

New York State stands out with 10,899 acclaimed deaths related to covid-19. The number of deaths in New York City is 35.36% of the whole nation. (Something strange is going on. Today, suddenly, thousands of addition covid-19 "probable" related cases were added to NYC without increase of other cities in NY at the same time.) President Trump mentioned this "in passing" that they were heart attacks "but 'probably' related to covid-19." Were these thousands of deaths added because more money would come to NYC? The petri dish is gathering many kinds of data beyond medical. The petri dish is full of bad stuff.

In the midst of the worldwide carnage, something good is happening in different areas, especially Africa, with halting infections of covid-19. We observed that the Africans where malaria is a problem have used quinine – now the people seem to resist the virus. I mentioned in a previous Lesson that a campaign would form that quinine does not prevent the virus. My charts indicate otherwise.

My charts of the African countries have not changed much in a day. It is evident that more people are being covid-19 tested. All of Africa has recorded a remarkably low total of 938 covid-19-related deaths as of today (4/15) including a total of only 7 deaths in the malaria troubled countries of Guinea, Botswana, Burundi, Zambia, and Malawi where

quinine seems to prevent the covid-19 from killing their people. One death yesterday (4/14) in Guinea.

TODAY'S WORLD REPORT 04/15/2020: Covid-19 confirmed 2,064,155 infected worldwide with 137,020 reported deaths. US confirmed 639,055 infected and 30,925 deaths.

What is Going On Behind the Chaos?
Science and Math are Used to Deceive the Public

Glycoscience NEWS Lesson #67 for 2020

04/16/2020

Covid-19 is the "tipping point" that has pushed thousands over the edge. Legitimately, coronavirus may be the cause for all the many deaths while the people actually died from other causes. Never was the flu to be fully blamed for all the deaths it "caused" – the flu virus was the "tipping point." Flu deaths have been attributed to other diseases, mostly pneumonia. If the truth be known, most of the covid-19 deaths have been pneumonia. We know lung and heart conditions are a factor.

The facts of science and math have been expressed deceptively. A lie has a twin brother, his name is Deceit. But, it is science and math that provide the data. I'm very excited to mine the data to see what facts it tells us about the big picture and many smaller pictures. Much good will come out of this worldwide event that even the chaos will help us understand. To know the "tipping points" of events within the big event will provide a wealth of valuable information capable of changing the planet for the good. However, the data must be mined with an integrative mind-set of *"what works best."*

You ask, *"What defines 'tipping point'?"* In defining *"tipping point"* examples, I lost my wife who complained that my explanation made her brain hurt. The fact is that an infinitesimal element of matter or energy can impact an environment.

In biology, one drop of contaminated blood added to all the blood in your body can kill you. In glycobiology, a 1/millionth of an unprotected area on the surface of a single cell can make that cell vulnerable and consequently damage the whole body.

Here is my description of a *"tipping point"*: Picture two buckets of water any size but equal on a balancing scale. The buckets are perfectly balanced to the drop. Now, you choose which bucket in which you put one drop of water. The bucket you chose is now heavier. It slowly starts to lower on the scale. As the heavier bucket lowers – it overcomes the lighter bucket and with each measurable distance of lowering of the bucket, all the weight in the bucket contributes its weight and increases the falling speed. Over time, the heavier bucket of water will hit bottom where it crashes and burns. Well, it can't burn because it's water. But, you get the point.

The tipping point requires an infinitesimal amount of change to tilt the scale. So, in fact, in science, the butterfly effect becomes a reality. In quantum physics, science can explain how the butterfly's wings in South America can, indeed, contribute to a storm in North America (www.tinyurl.com/butterflywings).

Yes, cause and effect are infinitesimal but change is evident.

"You could not remove a single grain of sand from its place without thereby ... changing something throughout all parts of the immeasurable whole." So said Johann Gottlieb Fichte in *The Vocation of Man* (1800).
(https://tinyurl.com/TheVocationalMan)

Sensitive dependence of known and unknown conditions, as infinitesimal as they may be, are involved in the outcome of chaos theory or reality. Henri Poincaré (1890) proposed that such phenomena could be common in meteorology.

We all have two buckets of health – moment-by-moment we put stuff in one of the buckets with each breath we take, each drink we swallow, and each bit we eat. Our health did not get to where it is overnight – it was over time. Who said, *"If I knew I were going to live this long, I would have taken better care of myself!"*?

Science and math are both accurate when we apply them to the number of covid-19-related deaths. But, like the flu, most all of the deaths were probably caused by something else. The coronavirus was the "tipping point" that pushed the most vulnerable, the severely ill, over the edge. The healthy were hardly affected but could still infect others as a carrier. The people on this first wave were hurt because we were IGNORANT. As we learn and prepare, the second wave of covid-19 will mostly hurt those who NEGLECT.

African charts have not changed much. Covid-19 is low and as of today (4/16) the total remains only 7 deaths in the malaria-troubled countries of Guinea, Botswana, Burundi, Zambia, and Malawi where quinine seems to prevent the covid-19 deaths.

TODAY'S WORLD REPORT 04/16/2020: Covid-19 confirmed 2,158,594 infected worldwide with 145,533 reported deaths. US confirmed 671,331 infected and 33,284 deaths. New York City appears to be the epicenter with a reported 11,477 and 3,262 deaths in New York State outside of the big city.

Petri Dish Studies will Expose World Cultures

Glycoscience NEWS Lesson #68 for 2020

04/17/2020

Medical experts and political experts look into the petri dish and see a killer virus – it appears very bad.

Let us look through the other end of the microscope. The political culture, the medical culture, and the world cultures are the ones in the petri dish – and they appear very bad.

The politics and medical leaders have shined a light on the virus and the light is shining on the medical culture scrambling for answers.

In the days ahead the world will see if any of the models are correct and what works and what does not work. The political leaders will call the shots for their regions. Each governor and mayor are in their own petri dish for the world to see. Keep your eyes on New York City and New York State. Today the mayor of NYC announced that he has no expectations of opening the city at least until July or August.

Things are changing rapidly. The integrity is askew for a true count of people who have been tested positive and for the number of deaths. The mayor of NYC dumped into the count thousands of heart attack deaths that had not even been tested for the virus but he thought they might have had covid-19. It was also reported that China announced that their account of deaths may be off by 50% but their increased numbers are not yet in the world count.

We live in the Data Age and researchers have the opportunity to gather more worldwide data than anytime in history with the ability to mine this data from thousands of different views within the big picture, from millions of smaller snapshots, and from an incalculable number of perspectives.

The focus will remain on the health and well-being of our people. However, we cannot allow the cure to be worse than the virus. When freedom is removed, the fatality of a people and regions will pay a much bigger price. This is where gathering the data will be important but more important is, "will we accept the hand-writing on the wall?" Regions at play are different countries and how they have responded to the pandemic. NYC and New York State provide their own petri dish cultures.

Regardless of how rotten the culture appears – I am excited to see lies and deceit exposed and to see the culture of our nation and our medical system renewed to our nation's origin and the Hippocratic Oath of medicine of DO NO HARM.

African charts have not changed much. Covid-19 is low and as of today (4/17) the total is only 9 deaths in the malaria-troubled countries of Guinea, Botswana, Burundi, Zambia,

and Malawi where quinine seems to prevent covid-19 deaths and we do not know the underlying conditions.

Although the integrity of these figures are in question – TODAY'S WORLD REPORT 04/17/2020: Covid-19 confirmed 2,244,303 infected worldwide with 154,219 reported deaths. US confirmed 702,164 infected and 37,054 deaths. New York City appears to be the epicenter with a reported 13,202 and 3,825 deaths in New York State outside of the big city.

A Snapshot of Coronavirus Damage in 50 States

Glycoscience NEWS Lesson #69 for 2020

4/18/2020 – HOUSTON –

This is a revealing study of the loss of life due (at least in part) to the coronavirus. The 50 states are ranked from lowest to highest loss per capita. The stats show that the number of deaths so far range from 0.2 to 51.5 people per 100,000. This snapshot is one petri dish observation.

Stats were recorded ~1st week to mid-April 2020

50. Wyoming	0.2 per 100,000
49. West Virginia	**0.5 per 100,000**
48. Utah	0.6 per 100,000
47. Hawaii	**0.6 per 100,000**
46. Montana	0.7 per 100,000

45. South Dakota	**0.7 per 100,000**
44. Alaska	0.8 per 100,000
43. North Carolina	**0.9 per 100,000**
42. Nebraska	1.0 per 100,000
41. Arkansas	**1.0 per 100,000**
40. Texas	1.1 per 100,000
39. Minnesota	**1.2 per 100,000**
38. North Dakota	1.2 per 100,000
37. Oregon	**1.3 per 100,000**
36. Maine	1.4 per 100,000
35. Iowa	**1.4 per 100,000**
34. New Mexico	1.5 per 100,000
33. Arizona	**1.7 per 100,000**
32. Virginia	1.7 per 100,000
31. South Carolina	**1.7 per 100,000**
30. Tennessee	1.7 per 100,000
29. New Hampshire	**1.7 per 100,000**
28. California	1.8 per 100,000
27. Idaho	**1.9 per 100,000**
26. Alabama	2.1 per 100,000
25. Missouri	**2.2 per 100,000**
24. Kansas	2.2 per 100,000
23. Florida	**2.3 per 100,000**
22. Ohio	2.3 per 100,000
21. Kentucky	**2.4 per 100,000**
20. Oklahoma	2.5 per 100,000
19. Wisconsin	**2.7 per 100,000**
18. Mississippi	3.3 per 100,000
17. Nevada	**3.8 per 100,000**
16. Delaware	4.2 per 100,000
15. Maryland	**4.3 per 100,000**
14. Pennsylvania	4.4 per 100,000
13. Vermont	**4.5 per 100,000**
12. Georgia	4.6 per 100,000
11. Indiana	**5.2 per 100,000**
10. Colorado	5.4 per 100,000

9. Illinois	6.3 per 100,000
8. Rhode Island	6.9 per 100,000
7. Washington	7.0 per 100,000
6. Massachusetts	12.2 per 100,000
5. Michigan	16.0 per 100,000
4. Connecticut	16.9 per 100,000
3. Louisiana	19.0 per 100,000
2. New Jersey	27.4 per 100,000
1. New York	51.5 per 100,000

African charts have not changed much but leaders there are very aware of the feared tinderbox possibilities. Covid-19 is low and as of today (4/18) the total remains at only 9 deaths in the malaria- troubled countries of Guinea, Botswana, Burundi, Zambia, and Malawi where quinine seems to prevent covid-19 deaths and we do not know the underlying conditions.

Report: 80-year-old Co-Founder of Teen Challenge, Don Wilkerson, recently recovered from covid-19 after he was given the quinine-based Hydroxychlororoquine.

TODAY'S WORLD REPORT 04/18/2020: Covid-19 confirmed 2,328,600 infected worldwide with 160,706 reported deaths. US confirmed 734,969 infected and 38,903 deaths. New York City appears to be the epicenter with a reported 13,157 (yesterday the report was 13,202 ?) and 4,545 deaths in New York State outside of the big city.

Explore the new frontier of medicine Glycoscience – learn how cells communicate. Download the *Glycoscience Whitepaper* at www.GlycoscienceWhitepaper.com

---

Some people are poised to exploit the pandemic in every creative way possible.

The Mafia and the Coronavirus

Glycoscience NEWS Lesson #70 for 2020

4/19/2020 - ITALY –

Question: Why do Bad Guys do bad things?

Answer: Because that is what Bad Guys do and they are good at it.

In Italy things have gone from bad to worse. From southern Italy, a report is that Mafia clans are providing everyday necessities in poor neighborhoods, offering credit to businesses on the verge of bankruptcy with plans to take a big part of the billions of euros scheduled as stimulus funds. Powerful Mafia families control the European cocaine market and are using the pandemic to advance their empire. The Mafia is deeply embedded in the economy and they are taking advantage of lowered restrictions of the supply chain because of the pandemic. They are infecting an already weak vulnerable economy with filthy lucre.

The invisible virus is destroying people and economies. In the spirit of death and destruction, thieves will pillage what they can from the top to the bottom. They are poised to exploit the pandemic in every creative way possible. Two cultures are at play. The **Good Guys**: (Everyone wants to be seen as the **Good Guy**.) Neighbor helping neighbor, expecting nothing in return. Then there are the **Bad Guys**: (Everyone can be a **Bad Guy** – because RATIONALIZATION is the greatest of all sins – RATIONALIZATION is the bag in which we carry all the other wrong doings.) The **Bad Guys** take advantage by killing, stealing, and destroying – taking for themselves. The **Bad Guys** put on white hats to make it look like they are the **Good Guys**.

Around the world, the very structure of government funding is designed to incentivize corruption and to bring the **Good Guys** alongside the **Bad Guys** as accomplice. Why did the mayor of New York City toss heart attack deaths in with coronavirus deaths? Why do so many **Good** doctors prescribe chemotherapy for so many of their patients? Shush!

We err only 3 ways: IGNORANCE, NEGLECT, and REBELLION.

I recommend a very "white hat" move for the use of filthy lucre that will serve as an act of kindness and open the door to help an untold number of people find medical innovations in Glycoscience. Go to the **Texas Endowment for Medical Research, Inc.** website, a 501(c)(3) non-profit charitable organization.

At www.TexasEndowment.org, click on "contribute" and give as much money as it takes for you to feel good about helping those who need help more than you. You may make a designated gift to assist specific individuals, specific research, help Africa, or children (contributions for children can be earmarked for *"Listen to the Cries of the Children*."

- -

LOL - A Good Laugh is Good Medicine: Bank lobbies are closed in Texas lockdown but will soon reopen. We cannot wear sun glasses into banks – but we may soon be required to wear masks. I have to chuckle when I see a giant bank vault wide open and the pens chained down.

- -

My African charts indicate more testing and more testing positive in several countries there. With 1,124 confirmed coronavirus deaths in all of Africa – the continent remains a tinderbox with leaders fearful of the potential. As of today (4/19) there are only reported 11 deaths in the malaria-troubled countries of Guinea, Botswana, Burundi, Zambia, and Malawi where quinine may be preventing the spread.

TODAY'S WORLD REPORT 04/19/2020: Covid-19 confirmed 2,404,249 infected worldwide with 165,234 reported deaths. US confirmed 759,687 infected and 40,682 deaths. New York City appears to be the epicenter with a reported 14,451 and 3,734 (yesterday number was 4,545) deaths in New York State outside of the big city. Italy reports 178,972 confirmed infected with 23,660 deaths.

The Petri Dish Holds the Answers

**It's not about who is Right or Wrong –
It is about Who is Right and Who is Left**

Glycoscience NEWS Lesson #71 for 2020

04/21/2020 – HOUSTON –

This is one heck of a petri dish test coming up! Let's look at it closely. The culture is changing. Will the culture be determined by the medical establishment, the politicians, or the people? New data suggests lockdowns may be futile and may even cause more deaths. WOW, we need to discover WHAT IS THE BEST PATHWAY to deal with this invisible enemy and NOW!

Here's the test that will settle the debate to lockdown or not to lockdown. Which saves the most lives and the economy?

(A) Lockdown; (B) No Lockdown.

Is this a medical decision or a political decision? What if it were not necessary to shut down the US or world economy? What if our economy is destroyed by the wrong decision that also results in more deaths? Is this whole chaos thing politically driven? Before we jump to such a *"totally ridiculous assumption"* – let's look at the FACTS.

The petri dish will reveal the FACTS about lockdown. The picture is developing, the culture is aging, so we can graphically compare the two choices – to lockdown or not to lockdown.

Here are **2 FACTS** unknown a few days ago: **FACT #1** Coronavirus is much more infectious than originally thought. **FACT #2** Coronavirus is less dangerous than originally thought.

These **2 FACTS** mean that the number of those infected is massive – perhaps more like the flu. It appears LESS dangerous than the flu – *unless the immune system is already compromised*. The cry is for massive testing. Massive testing is coming. The numbers will prove covid-19 is less dangerous than thought. The measuring tool is the ratio of deaths caused by covid-19 to population.

Were Ventilators another Wrong Choice?
The other day, the cry was for ventilators, *"We need thousands of ventilators!"* For the price of a new car, ventilators were delivered. New studies indicate that 80% of those put on a ventilator DIE because the patient does not get enough oxygen.

A hyperbaric chamber is a more effective choice. Oxygen therapy works. Lowering oxygen intake kills.

OPPORTUNITY FOR ALLOPATHIC MEDICINE This is the opportunity for the allopathic medicine culture to give it their best shot. And, that is exactly what they are doing and the results are coming in.

We are entering the age of INTEGRATIVE MEDICINE. Let US (the people) require integrating WHAT WORKS alongside the best the traditional medicine has to offer. The sad truth is that most doctors and healthcare professionals have not been taught immunology much less NATURAL IMMUNOLOGY. Glycoscience is the bedrock of immunology because it deals with the very base of glycans, glycoproteins, and glycolipids. We have learned HOW TO MANIFEST IMPROVEMENT through glycosylation – that means, we know how to improve the quantity and quality of these glycan in the human body to modulate the immune system.

Regardless of how rotten the culture appears – I am excited to see lies and deceit exposed and to see the culture of our nation and our medical system renewed to our nation's origin and the Hippocratic Oath of medicine of DO NO HARM.

African charts are nearly unchanged. Covid-19. Today (4/21) 13 deaths are reported in malaria-troubled countries of Guinea, Botswana, Burundi, Zambia, and Malawi where quinine seems to prevent the covid-19 deaths.

TODAY'S WORLD REPORT 04/21/2020: Covid-19 confirmed 2,544,769 infected worldwide with 175,621 reported deaths. US confirmed 814,587 infected with 43,796 deaths. New York City is the epicenter with a reported 14,604 deaths and New York State outside of the big city reported 4,102 deaths.

Chaos Goes Viral

Chaos happens when you are unable to prioritize a solution

Glycoscience NEWS Lesson #72 for 2020

04/22/2020 – HOUSTON –

CORRECTION: Typo in yesterday's Lesson about ventilators. The word should have been: hyperbaric

Were Ventilators Another Wrong Choice?
The other day, the cry was for ventilators, *"We need thousands of ventilators!"* For the price of a new car, ventilators were delivered. New studies indicate that 80% of those put on a ventilator DIE because the patient does not get enough oxygen. A hyperbaric chamber may be more effective. Oxygen therapy works. Lowering oxygen in the blood kills.

- - - - - - - - - - - - - - - - -

Chaos Goes Viral but appearance is subdued. In the next days much will be revealed. Answers coming from the petri dish will surprise you.

Expanded Testing –
Integrity of the Tests are in Question

Testing for the coronavirus is expanding and raise additional questions. Reports confirm that many individuals are asymptomatic. One report of 200 randomly tested people showed 60 tested positive for covid-19. Now, we have autopsies testing positive which has prompted claims that there were earlier fatalities.

How many cases have been missed?

Forensic science will soon reveal major mysteries. Autopsies of people reportedly died in California on February 6th and 17th, both tested positive contrary to previous thought that the 1st covid-19 death in the US occurred on 2/29/2020, in Kirkland, Washington. Forensic work is also moving back the date of events in Wuhan, China. How long has the virus been with US? What will we learn from mass testing? It will be exciting to discover new evidence for many unanswered questions.

The coronavirus models have failed repeatedly. China and US are fighting an information war. Facebook hired a "fact checker" who worked in the lab in Wuhan. The World Health Organization has been discredited. The Media fuel the flames with mis-information. The petri dish is revealing corruption and biases in the culture.

Ignorance, neglect, and bad intent makes for mass confusion with the invisible enemy called, "*coronavirus.*" Or is it "*covid-19?*" No, it is, "*SARS-Cov-2.*" Politicians, CDC, and WHO are confused - and many disqualified and await to hear the words, "*You are fired!*"

The prescription from the CDC is "*get your flu shot*" (that is not very affective). Medical scientists still believe allopathic medicine is the answer and consider the word "*alternative*" as a fighting conversation.

The word "*complimentary*" is not as combative as "*alternative.*" But when we use the word "*complimentary*" it is often received as a condescending spirit of superiority.

Integrative Medicine is where we are able to come along side the doctor and provide a pathway that will benefit the patient and make the doctor more successful and respected.

How many presumed covid-19 deaths were from the flu, pneumonia, or other complications? Many people who got the flu, died from pneumonia? Probably, most people who got covid-19, died from pneumonia or another illness.

This is allopathic medicine's opportunity to solve this worldwide pandemic. It is time to give it their best shot. They are counting on a vaccine. But...

The **Age of Integrative Medicine** is now. We are educating doctors and healthcare professionals in Glycoscience. Glycoscience is the bedrock of immunology because it deals with the very base of glycans, glycoproteins, and glycolipids. Glycosylation will manifest improvement of the immune system – that means, we know how to improve the quantity and quality of these glycans in the human body to modulate the immune system.

African charts are nearly unchanged. Covid-19. Today (4/22) 14 deaths reported in the malaria-troubled countries of Guinea, Botswana, Burundi, Zambia, and Malawi – quinine seems to prevent the spread.

RTODAY'S WORLD REPORT 04/22/2020: Covid-19 confirmed 2,628,527 infected worldwide with 183,424 deaths. US confirmed 842,376 infected with 46,688 deaths. NYC epicenter with a reported 15,074 deaths and NYS outside NYC 4,290 deaths. – Total US tests 4,466,559.

Announcement: Expect us to reveal a startling discovery about coronavirus – covid-19 – SARS-CoV-2 in a few days.

Houston Judge Wins First Petri Dish Award

Glycoscience NEWS Lesson #73 for 2020

04/23/2020 – HOUSTON

Harris County Judge Lina Hidalgo wins the first **Petri Dish Award**. The recently elected inexperienced judge strutted her authority with an order requiring everyone over age 10 to wear face masks for 30 days (with little exceptions) or be fined $1,000. A well-known medical doctor immediately sued to block her order that is more restrictive than orders from the governor of Texas or the President of the United States. Police officers, sheriff deputies, and the mayor of Houston announced that her edict would not be recognized. The outcry from the public and President reached way over the requirement for the judge to receive the **Petri Dish Award**. It is requested that she resend her preposterous overreach order before it is to go into effect Monday.

The **Petri Dish Award** is chosen by the people's disagreement and outspoken objection to common sense. The petri dish represents a small shallow transparent testing dish used to examine a growing culture to determine the toxic danger to the community. We welcome nominations for future **Petri Dish Award** recipients.

Looking through the other end of the microscope
Medical experts and political experts are looking into the petri dish and examining the killer virus to learn how it works and how to respond – it appears very bad. At the same time, the coronavirus provides us the opportunity to look through the other end of the microscope. It is the political culture, the

medical culture, and the world cultures who are in the petri dish for US to examine – and they often appear very toxic.

The politics and medical leaders have shined a light on the virus and the light is shining on the medical culture scrambling for answers. The petri dish allows us to examine the cultures used and to determine what saves the most people and that which kills the most people. The results will be clearly exposed in the petri dish.

In the days ahead the world will see if any of the models are correct and what works and what does not work. The political leaders will call the shots for their regions. Each governor, mayor, and judge are in their own petri dish for the world to see. Tough decisions are to be made – to lockdown or not to lockdown. Every move of the decision makers show up in the petri dish.

The Data Age in which we live is accumulating the information. Researchers have the opportunity to mine the data from thousands of different views within the big picture, from millions of smaller snapshots, and from an incalculable number of perspectives. Oh, the stories to be told – the lessons we will learn!

Keep your eyes on New York City and New York State. The mayor of NYC announced that he has no expectations of opening the city at least until July or August. Cause and effect brings the petri dish alive with activity.

Regardless of how rotten the culture appears in the petri dish – I am excited to see lies and deceit exposed and to see the culture of our nation and our medical system renewed to our nation's origin and to that of our Father of Medicine and his Hippocratic Oath to DO NO HARM.

My African charts indicate more testing and more testing positive in several countries there. The continent remains a tinderbox with leaders fearful of the potential. As of today (4/23) there are only reported 14 deaths in the malaria-troubled countries of Guinea, Botswana, Burundi, Zambia, and Malawi where quinine may be preventing the spread.

TODAY'S WORLD REPORT 04/23/2020: Covid-19 confirmed 2,708,885 infected worldwide with 190,858 reported deaths. US confirmed 869,170 infected and 49,954 deaths. New York City appears to be the epicenter with a reported 16,388 deaths plus 4,453 deaths in New York State outside of the big city. Reported tests conducted in the US is 4,660,250.

Ventilators are not Helping!
**Covid-19 - Coronavirus - SARS-CoV-2
is Different than any other virus**

Glycoscience NEWS Lesson #74 for 2020

04/24/2020 – HOUSTON –

Update: The decree issued by the Harris County judge goes into effect Monday that everyone in Houston who does not wear a face mask will be fined $1,000. But, because of opposition, the mayor has decreed that instead of a $1,000 fine, the police will give the "criminal" a free face mask.

- - - - - - - - - - - - - -

China Admits Many More Died than Reported
A new study claims it may be 10 times the earlier report. China figures times 5 or 10 would be more accurate says economics

professor Lucia Dunn of Ohio State University, and pathology professor Mai He of Washington University School of Medicine, says in a research paper published this week.

Massive Testing will be Dynamically Beneficial

The massive testing that is beginning will show us a whole new picture that I believe will be dynamically beneficial. With improved testing integrity, we will probably learn that millions of people test positive that had no idea they had covid-19. This will provide a great advantage to better understand the invisible enemy. More data will verify that covid-19 is much more contagious than the flu and much less deadly – except for those with compromised immune systems.

New Theory About Covid-19
Covid-19 is different than other viruses.

New evidence contradicts yesterday's knowledge. It is a deceitful invisible enemy. Scientists are forming new theories causing even more chaos and confusion because it contrasts with current thought. Covid-19 causes problems with the respiratory system – especially the lower region of the lungs. For this reason, the cry went out for more ventilators – but 80% to 88% of the people put on ventilators die. So, what's going on?

When an individual has a primed immune system, covid-19 simply runs its course – problem resolved.

However, when a weakened body allows covid-19 to progress to a critical level, the patient spirals into a state of hypoxemia – where the body cannot provide enough oxygen to the body. Obviously, an inadequate oxygen supply starves the organs and they shut down rather quickly, especially the brain which requires 25% of the total oxygen intake. The ventilators are worthless – maybe more than worthless. They may be killing people.

The ventilator may be getting oxygen to the lungs but the required level is not getting into the blood. Perhaps adding a hyperberic chamber would help but something is horribly wrong and people are dying. What is the problem? Covid-19 is doing something very differently.

Patients have life-threatening low oxygen levels in the blood; however, doctors have observed they do not show the normal distress signs expected with lung dysfunction. The doctor scratches his head. Yes, the covid-19 is indeed different. There is something that is causing this. What is it? We will discover a possible answer tomorrow!

My African charts indicate more testing and more testing positive in several countries there. The continent remains a tinderbox with leaders fearful of the potential. As of today (4/24) there are still only reported 14 deaths in the malaria-troubled countries of Guinea, Botswana, Burundi, Zambia, and Malawi where quinine may be preventing the spread.

TODAY'S WORLD REPORT 04/24/2020 5:30 PM: Covid-19 confirmed 2,790,786 infected worldwide with 195,920 reported deaths. US confirmed 890,524 infected and 51,017 deaths. New York City appears to be the epicenter with a reported 16,646 deaths plus 4,512 deaths in New York State outside of the big city. Reported tests conducted in the US is 4,692,797.

World Health Organization Model is Wrong and Here's Why

Glycoscience NEWS Lesson #75 for 2020

04/25/2020 – HOUSTON

It is my observation and calculation that the death ratio in America for coronavirus will settle out to be ~0.2 instead of the astronomical fluctuating figures of the President's Task Force. I arrived at that estimation taking into consideration 3 FACTS: FACT #1: the infection rate is greater than the flu; FACT #2: the infection is less deadly than the flu (except for those with a compromised immune system); and FACT #3: herd immunization that will soon burn out the spread.

Dr. David Atlas, at Stanford's Hoover Institution, agrees and adds that the **"*overwhelming majority of people do not have any significant risk of dying from covid-19.*"** He believes herd immunization (he calls it, "*vital population immunity*") is prevented by the isolation policy which prolongs the problem.

Dr. Atlas outlined that a recent Stanford University study estimates the fatality rate for infection with the coronavirus is likely 0.1% to 0.2%. The isolation policy of the WHO estimates 3.4%. Dr. Atlas and I agree that the isolation policy is the wrong model in comparison of herd immunization which may soon be proven (by my petri dish analogy) for covid-19. This is further substantiated by New York City figures: individuals 18 to 45 years old in New York City (which represents the age group for 1/3 of all US deaths) attributed to the virus = 0.01% or 11 people per 100,000 population. Individuals less than 18 years old, the death ratio = 0.00% or zero per 100,000.

The Power of Asymptomatic Individuals

Dr. Atlas uses New York City as an example that shows hospitialization figures from covid-19 for those less than 18 years of age is 0.01%. For individuals 65 to 74 years of age who were hospitalized represent only 1.7%. Dr. Atlas notes that medical science has proven over decades that infection itself allows people to generate an immune response, thereby controlling the spread within the population. Herd immunization of other viral diseases is more effective than Big Pharma's best shots. Shots are called vaccines and never called immunization because they are not. Medical care is not necessary for the most coronavirus-infected people. Dr. Atlas said, "*It is so mild that half of infected people are asymptomatic, shown in early data from the Diamond Princess ship, and then in Iceland and Italy.* [Asymptomatic people] are "*falsely portrayed as a problem requiring mass isolation,*" he said.

"*In fact, infected people without severe illness are the immediately available vehicle for establishing widespread immunity,*" Atlas wrote. "*By transmitting the virus to others in the low-risk group who then generate antibodies, they block the network of pathways toward the most vulnerable people, ultimately ending the threat.*"

New Theory about Covid-19 and
Why Ventilators are not solving the Problem

The oxygen absorption discrepancy demands that we solve a whole new mystery. Why is oxygen not being absorbed into the blood? It may not be the ventilator's fault. The very purpose of the ventilator is to move air in and out of the lungs. Oxygen attaches to the hemoglobin of red blood cells. The red blood cells carry oxygen through many miles of passages, delivering ~25% to the brain and the other 75% to organs and tissues of the body.

Does Covid-19 Damage the Hemoglobin?
Hemoglobin (a beta-1B-glycoprotein, the plasma protein) is responsible for binding and releasing oxygen molecules. Does covid-19 damage the hemoglobin and the glycoproteins involved? If so, a whole new approach is demanded TODAY! Stay tuned.

My African charts indicate more testing and more testing positive in several countries there. The continent remains a tinderbox with leaders fearful of the potential. As of today (4/25) there are still only reported 14 deaths in the malaria-troubled countries of Guinea, Botswana, Burundi, Zambia, and Malawi where quinine may be preventing the spread.

TODAY'S WORLD REPORT 04/25/2020 1 PM CT: Covid-19 confirmed 2,865,386 infected worldwide with 200,698 reported deaths. US confirmed 924,865 infected and 52,782 deaths. New York City appears to be the epicenter with a reported 16,853 deaths plus 4,743 deaths in New York State outside of the big city. Reported tests conducted in the US is 4,940,376.

Damaged Glycoprotein Plasma is HOW covid-19 is Killing People – Here's Proof!

Glycoscience NEWS Lesson #76 for 2020

04/26/2020 – HOUSTON –

We are learning new evidence about the invisible killer. Those who are open to integrative therapy can solve problems now as allopathic practices alone cannot act as quickly.

The malaria connection again

Reports are in that in the severe cases of covid-19 that lungs are damaged to the point of respiratory failure. There is no initial attack on the lungs that would effect their ability to pump air; rather, it is disruption of the hemoglobin to transport oxygen. *"Pulmonary edema is the most severe form of lung involvement."* This quote references both malaria and covid-19. The ventilators are getting oxygen to the lungs but absorption is critically low. The successful treatment of malaria, especially in Africa, has been with quinine for over a hundred years.

The compromised signals of the hemoglobin glycoprotein is the actual cause of death from the coronavirus. Hemoglobin is a plasma glycoprotein that binds, transports, and releases oxygen. This highly technical process uses iron as the oxygen-transport via the red blood cells from the lungs to the body. As needed, the hemoglobin releases the oxygen to power the metabolism and is able to transport carbon dioxide at the same time. This is why blood plasma is so important.

Facts we know and what we don't yet know:

1) Covid-19 is more infectious than the flu.
2) Covid-19 is less deadly than the flu to healthy individuals.
3) 80% to 88% patients put on ventilators reportedly die.
4) It is not the ventilator's fault.
5) Hemoglobin - Glycoprotein plasma signal failure is involved. Here are two possibilities. I believe it is (b).

      (a)   Covid-19 cause the signal malfunction.
          or
      (b)   Do the malfunction signals disrupt the immune system which permitted the coronavirus to attack the cell?

6) Proper oxygen absorption is inhibited in the blood stream.
7) Asymptomatic people can immunize the herd.
8) Improved Glycosylation – that is to improve the quantity and quality of glycans and glycoproteins on the surface of

    red blood cells – improves the cell signals and is a viable therapy for covid-19.

I hold in my hand, a published double-blind, randomized, placebo-controlled study verifying that my friends have actually increased the integrity of glycans and glycoproteins in human blood serum. The immunological and glycosylation changes manifest in the study were centered on marked changes in peripheral and intra-articular total immunoglobulin G (lgG) and Anti-citrullinated Protein Antibodies (ACPA) glycosylation. ACPA markers are postulated to have a pathogenic role in disease process.

Recent discovery: ACPA-IgG are extensively glycosylated in the variable (Fab) domain. Another study (not ours) indicates that more than 90% of ACPA-IgG molecules carry Fab glycans that are highly sialylated (referring to hemagglutination when viruses are mixed with blood cells, and the virus entry into cells of the upper respiratory tract). Another researcher: *"This molecular feature is striking and may provide a missing link in our understanding of the maturation of the ACPA immune response."*

Glycosylation is the area of research we have worked on for two decades and are praying for significant funding to benefit many through Glycoscience breakthroughs. View the red blood cell with it's glycans and glycoproteins at www.TexasEndowment.org

False Security and a Little Humor

False security may come from wearing a mask. Some were seen donning masks then having a long poker party, considering the mask a license to congregate. Remember when the casinos closed in China because of coronavirus? Then before long, they closed in Las Vegas. I saw a picture of bored gamblers sitting around the table playing poker using toilet paper rolls as chips. They said that the winner had a royal flush.

My African charts indicate more testing for covid-19 positive. 55 African countries confirm 32,296 infected – 1,426 reported deaths. The continent remains a tinderbox with leaders fearful of the potential. As of today (4/26) there are only 15 deaths reported in the malaria-troubled countries of Guinea, Botswana, Burundi, Zambia, and Malawi where quinine may be preventing the spread. 12 countries report 0 fatalities to covid-19.

TODAY'S WORLD REPORT 04/26/2020: Covid-19 confirmed 2,971,669 infected worldwide with 206,549 reported deaths. US confirmed 965,942 infected and 54,883 deaths. New York City appears to be the epicenter with a reported 17,280 deaths plus 4,712 deaths in New York State outside of the big city. Reported tests conducted in the US is 5,441,079.

Comparing New York and Hong Kong
2 cities in the petri dish – Let's examine the cultures

Glycoscience NEWS Lesson #77 for 2020

04/27/2020 – HONG KONG

This little study of how two large cities handle covid-19 is fascinating but there's a lot more to the story than this brief Lesson can report. Today (4/27) **NYC reports testing a total of 805,350 people. 292,000 are declared infected and 17,671 have died vs Hong Kong's report of 1,050 infections and 4 deaths.** What does this tell us? I'm not yet sure what it tells us but these two cases surely have a lot of data worth reviewing, mining, and applying to save lives.

When I was in Hong Kong, I remember my driver explaining apartment living. Several families live in the same tiny apartment and have communal bathrooms. Pointing over to my right, he said, *"About 5,000 people live in that apartment building."* I replied, *"That's the size of my home town when I was a boy."* Today, social distancing is not an option when multiple families are crowded together and rotate sleeping times. Sometimes a dozen people may live in cramped quarters. Small Hong Kong apartments are about the size of a New York jail cell. In New York (and Houston), criminals are freed so they can "social distance" at the peril of the general population. What is wrong with this picture? Someone is making some very wrong decisions for US. I was taught that making decisions is easy but to know the facts on which you base the decision is the difficult part.

Quantum Analysis of each Petri Dish Viral Event is to calculate into the equation limitless possibilities. We start with just a few obvious possibilities to help us understand the similarities and differences in New York and Hong Kong.

#1: Travel – Shortly after the Chinese Communist Regime became aware that the coronavirus virus was loosed, they stopped airplane and train travel out of Wuhan to cities in China. However, they did not stop flights out of Tianna (WUH) International Airport to other countries. There were direct flights to 18 countries. Some of the direct flights from WUH include New York City, San Francisco, Seattle, Rome, London, Paris, Tokyo, Osaka, Taipei, Kuala Lumpur, Sydney, Bangkok, Ho Chi Minh City, Seoul, Singapore, and Moscow. Connecting flights from WUH were for Canada, Central and South America, and Africa.

#2 City Population – Wuhan is a city of ~11 million people. New York City has ~8.4 million.

3# Social distancing when in public – The "lockdown" in Hong Kong and New York appear quite close. HK ruled out a complete lockdown. Hong Kong Chief Executive Carrie Lam Cheng Yuet-ngor said that to order a complete lockdown is contrary to the WHO recommendations. WHO Director-General Tedros Adhanom Ghebreyesus, stressed that the WHO disapproves of imposing travel or trade restrictions on China. The Hong Kong government tightened quarantine measures for Hong Kong residents returning from Hubei province, the epicenter of the outbreak.

Today (4/27) New York City, the US epicenter of the coronavirus outbreak closes some streets to expand sidewalks and bike lanes to provide more social distancing as lockdown measures continue, mayor Bill de Blasio announced. Many of the 8.4 million residents live in small apartments, and officials are concerned residents will flout social distancing rules the longer lockdown rules continue. The Police Department says that the city does not have the resources to protect people on those streets from drivers. Only a few days ago the mayor disputed the idea of closing streets to traffic, saying he did not believe it would work.

Glycosylation is the area of research we have worked on for two decades and are praying for significant funding to benefit many through Glycoscience breakthroughs. View the red blood cell with it's glycans and glycoproteins at www.TexasEndowment.org

My African charts indicate more testing and more testing covid-19 positive. 55 African countries confirm 32,296 infected with 1,426 reported deaths. The continent remains a tinderbox with leaders fearful of the potential. As of yesterday (4/26) there are only 15 deaths reported in the malaria-troubled countries of Guinea, Botswana, Burundi, Zambia, and Malawi where quinine may be preventing the spread. 12 countries report 0 fatalities to covid-19.

TODAY'S WORLD REPORT 04/27/2020 5 PM CT: Covid-19 confirmed 3,034,801 infected worldwide with 210,551 reported deaths. US confirmed 983,848 infected and 55,906 deaths. New York City appears to be the epicenter with a reported 17,671 deaths plus 4,893 deaths in New York State outside of the big city. Reported tests conducted in the US is 5,593,495.

Forensic Science & the Petri Dish

New York – Hong Kong – the World

Glycoscience NEWS Lesson #78 for 2020

04/28/2020 – HOUSTON –

My petri dish analogy of looking through the other end of the microscope gives us an opportunity to conduct an innovative forensic investigation. This will help us determine truth by uncovering facts buried under mounds of barnyard fertilizer – facts that would remain covered but for forensic discovery.

Many investigations start with preconceived notions that take even the high-tech detective in the wrong direction. Forensic science is not bias nor is it based on preconceived notions. Those nonscientific factors and emotions cannot be calculated into the equation. The con artist may plant false evidence to deceive the world. Truth is what matters. Truth is all that matters. Truth cannot remain buried.

In Lesson #77, we explored New York and Hong Kong similarities and differences. **#1: was Travel; #2 was City Population; #3 was Social distancing when in public.** Today we apply **#4 and #5.**

#4 Power of Forensic Science – The world of forensic science was advanced by Dr. Clarence "Lush" Lushbaugh, M.D., PhD. He formulated the method to determine the time of death of a murder victim. His work is used in movies and on CSI and NCIS. He died before he could know the impact his research had on my life. It is upon his shoulders that the pioneers of Glycoscience stand. In 1952, he made a significant discovery for the US Atomic Energy Commission; but he was unable to know how to make it work over time. His project failed and he was defeated. However, some 35 years later, my friend, Dr. Bill McAnally, a pharmacologist and toxicologist, studied Lush's work and discovered the first biological functional sugar with medicinal benefits. You are welcome to enjoy my brief tribute to Lush that I gave him back in 2012 at http://forum.endowmentmed.org/index.php?topic=415.0.

#5 War of words – Nothing is as it appears and the blame game compounds chaos. We are in the middle of misinformation filled with persuasive babbles. The invisible enemy set a trap to capture US. It came to kill, steal, and destroy. It has already ravaged the planet and devastated the economies of nations.

The internet provides an opportunity for false information to... to go viral. Forensic science can dig through all the barnyard fertilizer and verify truth from fiction. Even the biased "*fact checkers*" are too often deceptive. Statistics do not lie but statisticians do.

Quantum Analysis of each Petri Dish Viral Event will calculate limitless possibilities into the equation. With adequate funding, we can drill deep into problems and apply **Quantum Glycoscience** for breakthroughs.

We are all in this together and pointing fingers and blaming others is not productive. Sometimes when we seek truth and truth shows up – we are surprised – "*THAT was not what I was expecting.*"

The petri dish is exposing corruption that will surprise and may even shock the world. The culture of politics, the culture of Big Pharma, the culture of education can be improved. The war of words is too often an omelette of lies and deception sprinkled with truth.

Forensic science cuts through all barnyard fertilizer and casts aside emotion and biases. Smothered in self-evidence, truth is delivered on a platter so you and I can make rightful decisions. When the findings are complete, the enemies will hang on gallows of their own design.

Glycosylation is the area of research we have worked on for two decades and are praying for significant funding to benefit many through Glycoscience breakthroughs. View the red blood cell with it's glycans and glycoproteins at www.TexasEndowment.org

My African charts indicate (4/26) more covid-19 test positive. 55 African countries confirm 32,296 infected with 1,426 reported deaths. The continent remains a tinderbox with leaders fearful of the potential. As of today (4/28) there are still only 15 deaths reported in the malaria-troubled countries of Guinea, Botswana, Burundi, Zambia, and Malawi where quinine may be preventing the spread. 12 countries report 0 fatalities to covid-19.

TODAY'S WORLD REPORT 04/28/2020 5PM CT: Covid-19 confirmed 3,110,219 infected worldwide - 216,808 reported deaths. US confirmed 1,010,717 infected & 58,365 deaths. NYC is the epicenter with a reported 17,682 deaths plus ?4,893? (discrepancy)* deaths in New York State outside of the big city. Reported tests conducted in the US is 5,776,829.

> ** Discrepancies in New York State and New York City are very suspect. Today, the number was LESS than the day before.*

**The stigma of a *"conspiracy theory"*
ends when it is no longer a theory**

NIH Funded the Lab in Wuhan

Glycoscience NEWS Lesson #79 for 2020

04/29/2020 – WASHINGTON DC –

National Institutes of Health (NIH)

Dr. Anthony Fauci, American physician and immunologist served as the director of the National Institute of Allergy and Infectious Diseases (NIAID) since 1984. During the last administration, Dr Fauci funded coronavirus research at the **Wuhan Institute of Virology China.**

Last Friday (4/24), the NIH learned that funding was immediately terminated. The Trump administration cut off all funding for the project and requested that NIH inform the sponsor of the study, EcoHealth Alliance, to whom they gave the money. This came to light last week after *"conspiracy theory"* reports linked NIH to the work of the Wuhan level 4 laboratory. According to Rudy Giuliani, the last administration must be held accountable for the $3.7 million that we (that's US) granted to China for coronavirus research. Several addition millions may have also been sent to China from US for the project that is killing US and the world.

I found a Beijing NEWS story dated January 4, 2018: *"China's first bio-safety level 4 lab – China has opened its first bio-safety level four laboratory, capable of conducting experiments with highly pathogenic microorganisms, according to the national health authority.*"

Records in Washington DC indicate that US officials were concerned about the Wuhan lab back in January 2018. Two "*sensitive but unclassified*" messages asked for assistance to help the lab tighten its safety protocols were sent to Washington from the US Embassy in Beijing following visits to the Wuhan lab by a US diplomat and a science diplomat at the US Embassy. It was noted that the Wuhan lab did not have appropriately trained technicians and investigators needed nor the ability to safely operate the high-containment laboratory.

- - - - - - - - - - - - - - - - -

The World Health Organization (WHO)

What does **Taiwan**, **Australia**, and **New Zealand** have in common? They ALL disregarded the advice of the WHO. It does not require major forensic analysis to compare other countries to **Taiwan**, **Australia**, and **New Zealand** to see a pattern developing. As of today (4/29) the reported deaths from covid-19 in **Italy** is 27,682. Had they NOT followed the WHO guidelines no telling how many lives would have been saved. These 3 countries ignored the WHO's rules and here is the count for all their covid-19 deaths to date:

> **Australia – 90 deaths;**
> **New Zealand – 19;**
> **Taiwan – 6;**
> **compared to Italy – 27,682**

"WHO, you are fired."

- - - - - - - - - - - - - - - - - -

We will conduct further reviews of countries pulling away from the WHO flawed guidance system. We will also study the countries that are integrating "what works" along side allopathic medicine. This will help us understand why some other countries have such a low death rate. This tiny invisible evil enemy did not intend to provide US with such a beautiful forensic picture to view and understand. Today the report from India confirms they have only lost 1,079 lives to covid-19. We will take a closer look at India later.

Is the Center for Disease Control Next?

The CDC has lied and deceived US for years. Much forensic evidence – some Freedom of Information Act (FOIA) requests – conclude that the CDC has intentionally put out false information, put US in great danger, and followed Big Pharm's play book.

Forensic science cuts through all barnyard fertilizer and casts aside emotion and biases. Smothered in self-evidence, truth is delivered on a platter so you and I can make rightful decisions. When the findings are complete, the enemies will hang on gallows of their own design.

My African charts indicate (4/26) more covid-19 test positive. 55 African countries confirm 32,296 infected with 1,426 reported deaths. The continent remains a tinderbox with leaders fearful of the potential. As of today (4/29) there are still only 15 deaths reported in the malaria-troubled countries of Guinea, Botswana, Burundi, Zambia, and Malawi where quinine may be preventing the spread. 12 African countries report 0 fatalities to covid-19.

TODAY'S WORLD REPORT 04/29/2020 5PM CT: Covid-19 confirmed 3,193,961 infected worldwide - 227,644 reported deaths. US confirmed 1,039,909 infected & 60,066 deaths. NYC is the epicenter with a reported 18,076 deaths plus 5,234 deaths in New York State outside of the big city. Reported tests conducted in the US is 6,026,175.

Quarantine has a Significant Meaning Especially for this period of time we are in

Glycoscience NEWS Lesson #80 for 2020

In January, at the beginning of the new year 2020, I envisioned the Trends, the Changes, and the Expectations and asked, "Will this year be like the Roaring 20s? Will the year provide 2020 Vision into the future?"

The Latin root meaning for quarantine is **40**. The number **40** will ring loudly in 2020. **40** has been a significant number through history. The number **40** locks in a period of time. **40** represents a time of change. The number **40** is preparation time for the coming change. Quarantine is on everyone's mind. **40** is a time of expectation. Something is in the air other than a tiny invisible killer virus. Something is about to awaken a lot of people.

As we move into 2020 so does the decade of greatest change and polarization the world has ever seen. We are so privileged to live in the most exciting time in all history. Medical breakthroughs and wonderful expectations are on one hand while "on the other hand" shocking disasters will occupy the news.

The coronavirus got the attention of the world and this is an important time in history. Other times in history when the number **40** had significance in waiting, preparation, testing, punishment, trial, or probation: It rained **40** days and nights; the Hebrews wandered in the wilderness **40** years; **40** starts a new chapter.

My African charts indicate (4/26) more covid-19 test positive. 55 African countries confirm 32,296 infected with 1,426 reported deaths. The continent remains a tinderbox with leaders fearful of the potential. As of (4/29) there are still only 15 deaths reported in the malaria-troubled countries of Guinea, Botswana, Burundi, Zambia, and Malawi where quinine may be preventing the spread. 12 African countries report 0 fatalities to covid-19.

TODAY'S WORLD REPORT 04/30/2020 5PM CT: Covid-19 confirmed 3,193,961 infected worldwide - 227,644 reported deaths. US confirmed 1,069,664 infected & 63,006 deaths. NYC is the epicenter with a reported 18,069 ?(less total than yesterday deaths plus 5,380 deaths in NY State outside of the big city. Reported tests conducted in the US is 6,231,182.

The NY Putrid Petri Dish

Glycoscience NEWS Lesson #81 for 2020

05/01/2020 –

New York's Death Angel Had Help

The death angel has not passed over New York. New York has had more than 27% of the covid-19 deaths reported in the whole US (NY 23,449 to 63,019). That is more than 10% of the reported deaths of the whole planet (23,449 to 233,998). Figures are as of 04/30/2020.

Density of the nation's biggest city is a factor but other significant factors have greater cause. The governor and mayor point to world travelers, especially from China and Italy, as the reason for the outbreak and the staggering size of the problem.

Chaos Rules in Nursing Home Deaths

The official report of confirmed and presumed covid-19 nursing home deaths for the state of New York as of 04/30/2020 is 3,688. The number of **New York Nursing Home Deaths** is nearly 16% of the number of all deaths in the state and nearly 6% of all the deaths in the US.

Promising to protect the elderly who are the most vulnerable, I heard Governor Andrew Cuomo say, *"My mother is not expendable and your mother is not expendable and our brothers and sisters are not expendable."* From the other side of the governor's mouth he mandated that nursing homes admit and readmit patients who test positive for covid-19. When questioned about his policy, his response was to play ignorant by saying, *"That's a good question. I don't know."*

Gov. Cuomo's covid-19 nursing home policy put to death an untold number of the elderly. People are angry at the needless and strange NY policy that proved fatal to so many. This looks more like euthanasia than a reckless decision.

The New York policy is especially strange since the first US outbreak was in a nursing home on the other side of the country in Kirkland, Washington where 13 deaths took place – now we have reports that at least 24 others have died there. The governor's action is a blatant example of a politician doing the opposite of what they are saying. Remember, we miss the mark either by ignorance, neglect, or rebellion. I don't know Governor Cuomo and I'm not sure how he missed the opportunity to save lives. It could be ignorance and/or neglect.

About 2 weeks ago an Italian study found 99% of covid-19 fatalities suffered from other health issues and the average age of death was 79.5 years. Congress designed financial incentives to treat covid-19 which may account for the number of patients and consequently for the number of deaths. There is mounting evidence that all the mistakes and chaos for treating covid-19 is not just ignorance or neglect. Decisions are made for personal monetary gain or for that of a state or city.

My African charts indicate 40,804 tested positive for covid-19 in 56 countries in the continent and 1,679 are reported to have died. Africa remains a tinderbox with leaders fearful. As of today (05/01) there are only 15 deaths reported in the 5 malaria-troubled countries of Guinea, Botswana, Burundi, Zambia, and Malawi where quinine may be preventing the spread.

Looking through the other end of the microscope, we observe scientific and political "experts" manipulating circumstances. The world observes their contradicting words and actions. It is by looking through the other end of the microscope that we view the accomplices of the invisible mass killer as they attempt to achieve their own agendas. The petri dish exposes their culture. The people are tired of the deception of politics, Big Pharma, and education.

TODAY'S WORLD REPORT 05/01/2020 3 PM CT: Covid-19 confirmed 3,329,740 infected worldwide with 237,647 reported deaths. US confirmed 1,094,640 infected and 64,324 deaths. New York City appears to be the epicenter with a reported 18,399 deaths plus 5,393 deaths in New York State outside of the big city. Reported tests conducted in the US is 6,322,198.

Exciting Battle of 2 Drugs

Glycoscience NEWS Lesson #82 for 2020

05/02/2020 –

In an apparent battle of 2 drugs, the FDA, yesterday (5/1) issued an emergency-use "authorization" for the antiviral drug Remdesivir. Some medical scientists are downplaying **hydroxychloroquine** seemingly as a retaliatory response to the President's downplaying of Remdesivir. What works best may play out in the weeks ahead for the world to see. This is going to be interesting to watch.

Big Pharma continues its attempts to *"improve"* what naturally works into billion dollar waterfall opportunities. We see a battle between 2 drugs, **hydroxychloroquine (a quinine-based drug backed by President Trump)** and **Remdesivir (backed by Dr. Anthony Fauci). Remdesivir's** connection to Big Pharma raises suspicion of their desire to make giant profits from the coronavirus. **Hydroxychloroquine, on the other hand, is already generically available for less money**.

Earlier this week (4/29), Dr. Anthony Fauci announced that he was concerned about a possible leak about the development of an experimental drug to treat covid-19. So, this fear compelled him to reveal data from the Oval Office before this information leaked out. (*The Leak: Bloomberg NEWS indicated that Remdesivir was not as effective as projected.*) Fauci explained that the drug is made by Gilead Sciences and is going through rigorous clinical trials for FDA approval. Dr. Fauci said, "*It was purely driven by ethical concerns.*"

I am all for ethical concerns unless they differ with moral concerns. Who is Gilead Sciences? Gilead is a California-based company, but China applied for the patent through the Institute of Virology in Wuhan for Remdesivir on January 21. The drug has a worldwide market potential worth untold billions. The Wuhan Institute and the Beijing Institute of Pharmacology and Toxicology research was published in Cell Research Journal. The China studies reportedly found that Remdesivir compound and the chloroquine malaria drug are both "*highly effective*" in the control of coronavirus infection.

The compounds in Remdesivir have been used separately in humans and have "*a safety track record and shown to be effective against various ailments.*" The researchers wrote, [Remdesivir] "*should be assessed in human patients suffering from the novel coronavirus disease.*"

Before the FDA approves a drug, it must pass the LD50 level test which means the Lethal Dose must kill 50% of the animals. The USDA says that some 25 million animals may be killed each year because of this archaic FDA LD50 Rule before a drug is allowed to be **"safe for humans."** This is the secret the FDA uses to keep all toxin-free solutions off the market. After reading the side effects of Remdesivir, I know it had no trouble killing 50% of the animals in the study. The LD50 level provides the calculation formula of toxicity to know how much of the drug humans would have before it kills them.

My book, *To Kill A Rat*, explains the archaic LD50 level rule and how the FDA needs to be changed for the benefit of the people instead of Big Pharma.

I welcome an opportunity to provide a placebo for testing it along side either or both of these drugs – hydroxychloroquine and/or Remdesivir. My placebo is toxin-free and does not meet the LD50 requirement for a drug but does qualify as a placebo sugar.

Forensic science finds truth that shows what works best and provides the way to cut through barnyard fertilizer, cast aside emotion and biases. Smothered in self-evidence, truth is delivered on a platter so you and I can make rightful decisions. When findings are complete, those who deceive US will hang on gallows of their own design.

NEW African update (5/02) 1,827 new (within ~24 hours) covid-19 positive tests in 29 countries bring the official count of 56 African countries to 34,123 infected with 1,725 reported covid-19-related deaths. 46 new deaths were reported (within ~24 hours). The continent remains a tinderbox with leaders fearful of the potential. As of today (5/02) there are still only 15 deaths reported in the malaria-troubled countries of **Guinea, Botswana, Burundi, Zambia, and Malawi** where quinine may be preventing the spread. 11 African countries out of the 56 countries report 0 fatalities to covid-19. There were 10 new positive cases in **Zambia** reporting a total of 119 with 3 deaths (no increase). **Malawi** reported 1 additional positive case for a total of 38 also with 3 deaths (no increase). Pray for Africa.

TODAY'S WORLD REPORT 05/02/2020 6PM CT: Covid-19 confirmed 3,419,184 infected worldwide - 243,355 reported deaths. US confirmed 1,127,712 infected & 66,075 deaths. NYC is the epicenter with a reported 18,491 deaths plus 5,515 deaths in New York State outside of the big city. Reported tests conducted in the US is 6,816,347.

16,000+ Nursing Home Deaths

Death is imminent when chaos rules, euthanasia is accepted and morality is shelved.

Glycoscience NEWS Lesson #83 for 2020

05/03/2020 –

From Washington to New York – Nursing home deaths make up ~25% of all covid-19-related deaths. Some states report nursing home deaths are running over 50% of state deaths. Last week, President Trump announced that the Federal Emergency Management Agency is sending protective supplies to all of the nation's nursing homes. Governors are calling for testing of all residents and staff of long-term care facilities.

The US has ~15,600 nursing homes and ~28,900 senior living facilities, according to the CDC figures (more than 600 in New York) from 2016. This makes up for more than 2 million who are under long-term care.

We will learn more about the nursing home disaster in the days ahead.

Let's look at US for a minute and examine our own culture as we cross this period of time together. We ARE all in this together and we buffer each other. Families can become closer or further apart – our choice.

Let's grow closer together in knowledge and peace and happiness and joy and love for and with each other. Don't be like the husband and father who took time off from the internet (because the web shut down) and spent time with his family. When a buddy asked him how did it go with the family – he answered, "*They seem to be good people.*" Or, (one more) the mother, being alone with her son, said, "*Jimmy, do you think I'm a bad mother?*" He said, "*My name is Johnny.*"

I became very angry when I saw the needless deaths caused by ignorance, neglect, and rebellion. When the governor of New York enforced a policy that demanded covid-19 positive patients to be placed in nursing homes with the ailing elderly who did not have the virus. I called it, "euthanasia." I called it "murder." Yes, I was angry.

I was angry when my own mother was about to be killed in a hospital years ago. She was going "*down hill fast*" because the doctor was making bad decisions. I called an ambulance that came to the hospital to transfer my mother to another hospital. The hospital staff went ballistic and told me, "*You cannot do that. You cannot move her. The doctor has not dismissed her.*" My response was, "*I fired the doctor, he does not work for us any more and she IS getting in that ambulance. And, I will release the hospital of all responsibility.*" Had I not taken authority, she would not have lived another 20 wonderful years to be with her children and grandchildren.

The Virus is an Excuse for Extremists

We shall brace against the radical extremists who will use the virus to bring harm to US with anger and rebellion and turn the culture more rotten than before. New York may burn and more people will die. Let's put out fires of hate and heal those who others would hurt. We have been deceived in many ways. One Big deception is that the coronavirus has not killed many people

– they died from pneumonia or other complications of a compromised immune system.

Forensic investigation is exposing the culture of such petrification that it would make anyone angry – but be at peace, a new day is almost here – but it may get worse before it gets better.

Forensic science finds truth that shows what works best and provides the way to cut through barnyard fertilizer, cast aside emotion and biases. Smothered in self-evidence, truth is delivered on a platter so you and I can make rightful decisions. When findings are complete, those who deceive US will hang on gallows of their own design.

African Report (5/02) 1,827 new (within ~24 hours) covid-19 positive tests in 29 countries bring the official count of 56 African countries to 34,123 infected with 1,725 reported covid-19-related deaths. 46 new deaths were reported (within ~24 hours). The continent remains a tinderbox with leaders fearful of the potential. As of today (5/02) there are still only 15 deaths reported in the malaria-troubled countries of **Guinea, Botswana, Burundi, Zambia, and Malawi** where quinine may be preventing the spread. 11 African countries out of the 56 countries report 0 fatalities to covid-19. There were 10 new positive cases in **Zambia** reporting a total of 119 with 3 death (no increase). **Malawi** reported 1 additional positive case for a total of 38 also with 3 deaths (no increase). Pray for Africa.

TODAY'S WORLD REPORT 05/03/2020: Covid-19 confirmed 3,506,924 infected worldwide - 247,473 reported deaths. US confirmed 1,158,041 infected & 67,682 deaths. NYC is the epicenter with a reported 18,925 deaths plus 5,643 deaths in New York State outside of the big city for a total for NY of 24,568. Reported tests conducted in the US is 7,053,366.

Looking Beyond Vaccines to Antibodies

Glycoscience Holds the Secret for the Elusive Antibodies ● How to Proliferate Antibodies in your own body ●

Glycoscience NEWS Lesson #84 for 2020

05/04/2020 – Today's NEWS from Switzerland and Italy

Switzerland's search for antibodies

The Swiss School of Public Health is conducting a national study via network of 12 universities. The aim of the study called *Corona Immunitas* is to determine the level of immunity in people infected with covid-19.

Switzerland's Federal Office of Public Health announced it supports the research based on a proposal by Professor Milo Puhan, Head of the Institute of Epidemiology at the University of Zurich. The study seeks to determine the level of the human body's response to infection by measuring the antibodies in the blood. The first tests started in Geneva on 6 April 2020. Finding and using a reliable test is critical to target specific antibodies needed to immunize the body.

The Swiss findings will provide valuable information to help the world better decide how to respond to the pandemic. The idea is designed to guide decisions and eventual to develop a vaccination program. But...

We Can Go Beyond Vaccines

The findings from the Swiss study is significant in helping us go beyond vaccines to simply monitor the antibodies needed and to develop the proliferation plan for those specific antibodies through Glycoscience.

Meanwhile in Italy

Europeans and Americans are hoping antibodies is their answer to immunity to the coronavirus so they can reopen societies and economies. Politicians in Italy, the epicenter of Europe, are proposing issuing licenses to those who had beaten the virus and developed the right antibodies. Researchers and politicians worldwide have latched onto the idea that antibodies may be the solution to covid-19.

But many researchers are refusing to cooperate. Big Pharma is not looking for a simple solution. Big Pharma sees Trillions of Dollars in vaccine sales. It is interesting to see the back peddling, *"The promise of antibodies may not be what people have imagined. At least for now."* However, Italy's government will begin testing 150,000 people for antibodies as a *"sample survey on the spread of the infection."* Mario Plebani, coordinator of the antibody testing program is quoted, *"Immunity licenses are just rubbish."*

In a future Lesson, I will get more into the specifics of HOW we can proliferate antibodies and stem cells in the human body.

Forensic science finds truth that shows what works best and provides the way to cut through barnyard fertilizer, cast aside emotion and biases. Smothered in self-evidence, truth is delivered on a platter so you and I can make rightful decisions. When findings are complete, those who deceive US will hang on gallows of their own design.

African Report (5/04) 12,226 new test+ (within ~48 hours) covid-19 cases in 38 countries brings the official count of 56 African countries to 46,487 infected with 1,827 reported covid-19-related deaths. 96 new deaths were reported (within ~48 hours). The continent remains a tinderbox with leaders fearful of the potential. *As of today (5/04) there are 17 deaths reported (Guinea reported 2 deaths in the last 48 hours) in the malaria-troubled countries of **Guinea (9 deaths - 1,710 test+), Botswana (1 death - 23 test+), Burundi (1 death - 15 test+), Zambia (3 deaths - 137 test+), and Malawi (3 deaths - 41 test+).** 11 African countries out of the 56 countries report 0 fatalities to covid-19.* Pray for Africa.

TODAY'S WORLD REPORT 05/04/2020 3:30 PM CT: Covid-19 confirmed 3,562,919 infected worldwide - 249,712 reported deaths. US confirmed 1,172,670 infected & 68,326 deaths. NYC is the epicenter with a reported 19,057 deaths plus 5,647 deaths in New York State outside of the big city for a total for NY of 24,704. Reported tests conducted in the US is 7,123,222.

New FDA Rules on Antibody Tests for covid-19

Glycoscience NEWS Lesson #85 for 2020

05/05/2020 –

Well, that didn't take long. Yesterday, I reported on developments for antibody testing for covid-19 that could save an untold number of lives. Before yesterday was over, the FDA made a new ruling on antibody testing by tightening existing rules. The FDA concerns are in regard for the test's accuracy and reliability. Those who have recovered are asked to donate blood plasma for an experimental treatment that aims to transfer immunity to other patients. Testing antibodies may produce a strategy to identify who is safe and who is not safe.

The new rules require companies to submit within 10 days the required emergency-use paperwork and data proving that their tests work. The tests are designed to determine if patients have antibodies in their blood that make them more immune to the coronavirus. The FDA said they have authorized 12 tests and more than 200 are in pre-approval phases. There is a public concern that the FDA could erase some civil liberties.

Breakthroughs are coming – I am very excited about the antibody testing to determine quantity and quality of antibodies in the blood. In Glycoscience, we now know how to improve the quantity and quality of the glycans and glycoproteins naturally to produce the antibodies needed to shield the cells from covid-19 or any other virus. Through applied Glycoscience, we have learned how to improve glycosylation of the glycans, glycoproteins, and glycolipids to modulate the immune system.

FDA Protects US from Wrongdoers and I'm Grateful
We cannot make fraudulent claims and we should all be thankful for that fact. I wish I could report all the testimonies that I have received from many people who have had remarkable health benefits from the T/C+ that I formulated a few years ago. I have collaborated with several universities internationally and assisted in their research especially in the area of neurological and inflammatory challenges. I lean heavily on scientific evidence through research.

Improved T/C+ with X5 – Doctors, healthcare professionals, and the general public are welcome to help us help your family and friends participate in a Pilot Survey. **A natural ingredient in T/C+ with X5 is believed to destroy viruses** but may not be conclusive. Needs more research.

I have improved the formulation of T/C+ which is now called **T/C+ with X5** and it is designed to lower high blood pressure, aid brain function, reduce cell stress, serve as an anti-depressant, lower LDL cholesterol, sustain energy, aid stress tolerance, lower triglycerides, reduce arthritis, provide cardiovascular support, aid metabolic syndrome, inhibit fat cell enlargement, help protein folding, stabilize proteins, reduce inflammation, strengthen cell membrane, help overcome diabetes 1 & 2, nourish neurological processes, inhibit progression of Type 2 Diabetes, delay neurological malfunction, inhibit progression of beta-amyloid 40 and 42 which attributes to Alzheimer's.

We welcome individuals with the following health challenges into a **T/C+ with X5 – Six Month Pilot Survey** because research shows that **T/C+ with X5 may provide a higher quality of life and delay the onset of and sometimes may reverse certain neurological processes which include:** muscular dystrophy, MS, Parkinson's, Alzheimer's, ALS, stroke, spina bifida, sleep apnea, shingles, peripheral neuropathy, migraines, meningitis, Lyme's, fibromyalgia, epilepsy, ADD, encephalitis, Down's syndrome, dementia, AdHD, chronic fatigue, bipolar, autism, alcoholism, and Huntington's.

Multiple University Research Studies in various countries verify the functional components used in **T/C+ with X5** have significant health benefits. Evidence of neither synergy nor counter-synergy effect is determined. Health benefits are unique to individuals and others may or may not receive similar benefits. In our research, we continue to have slow-responders, hyper-responders and non-responders. Our objective is to discover ways and means to continue to improve better health

and to provide solutions with a DO NO HARM Safety policy in a drug-free environment.
For more information: JCSpencer@TexasEndowment.org

African Report postponed – will continue

TODAY'S WORLD REPORT 05/05/2020 3 PM CT: Covid-19 confirmed 3,640,835 infected worldwide with 255,096 deaths. US confirmed 1,199,238 infected & 70,646 deaths. NY is the epicenter with a reported 25,073 deaths (More Chaos in NY: 1,700 past deaths in nursing home claimed to be unreported. Reported tests conducted in the US is 7,285,178.

To Mask or Not to Mask?

Glycoscience NEWS Lesson #86 for 2020

05/06/2020 –

I still chuckle about the sign at the bank that tells me to take off my sunglasses – then requests I put on a mask.

To Mask or Not to Mask?
To mask or not to mask is but one of the confusing issues. Confusing directions came a few weeks ago when Dr. Fauci said masks were not necessary except for medical workers. To wear a mask may give a false sense of security.

New evidence shows that when I wear a mask, it may not protect me but it may protect others from me – because the infectious mist may go 12 feet without my mask and with my mask on the droplets may only go about 3 feet.

How judges and mayors handled the mask situation in Chicago and Houston is classic. Common sense is uncommon. But they don't hold a candle to what happened in Michigan when a security guard was shot to death attempting to enforce state policy to wear a face mask.

Hundreds of protesters descended on the Michigan Capitol building to demonstrate against the governor's stay-at-home order. Most of the protesters refused to wear a mask. Some people need to protest and then there are those who just like to protest and bring out their swastikas and nooses. We need to heed the expertise and those of higher learning – the problem is that they are hard to find. The chaos and harm caused by some injunctions put others at risk and can quickly outweigh the treats of the virus.

Safety and health are at the top of the list but not at the destruction of our God-given rights. I remember the battle of helmets for motorcycle riders. It makes sense to have on a helmet but the story goes that this motorcyclist was so opposed to the government telling him what to do that he was killed on his way to a protest rally because he wasn't wearing a helmet. I still support him in his protest.

We are learning so much from this invisible killer that we will never be the same. The world will never be the same.

I'm excited and perplexed at what else we are about to learn from Russia, Africa, New York, and various regions. Among the comforting things we will learn about covid-19 is that it may not be as dangerous as driving to work.

African Report (5/06) 4,767 new test+ (within ~48 hours) covid-19 cases in 35 countries brings the official count of 56 African countries to 51,254 infected with 1,960 reported covid-19-related deaths. 133 new deaths were reported (within ~48 hours). The continent remains a tinderbox with leaders fearful of the potential. *As of today (5/06) there are 20 deaths reported (Guinea reported 2 deaths in the last 48 hours) in the malaria-troubled countries of **Guinea (11 deaths - 1,856 test+), Botswana (1 death - 23 test+), Burundi (1 death - 15 test+), Zambia (4 deaths - 146 test+), and Malawi (3 deaths - 43 test+).** 11 African countries out of the 56 countries report 0 fatalities to covid-19.* Pray for Africa.

TODAY'S WORLD REPORT 05/06/2020 5:30 PM CT: Covid-19 confirmed 3,742,665 infected worldwide with 262,709 deaths. US confirmed 1,223,419 infected & 72,812 deaths. NY is the epicenter with a reported 25,231 deaths. Reported tests conducted in the US is 7,759,771. 189,910 Reported Recovered.

When Experts Get It Wrong, People Die

The "Guidance Plan" that health experts are using to guide US is flawed

Glycoscience NEWS Lesson #87 for 2020

05/07/2020

Professional guidelines and projections have been wrong repeatedly. The evidence verifies conclusively that WHO, CDC,

NIH, and FDA have been misguided by the same people attempting to misguide US. Their plan is filled with errors caused by ignorance, neglect, and deception.

Academic experts from Arizona State University and University of Arizoressna were guiding Arizona's Department of Health Services – until Monday evening (5/4) just before President Trump landed. Governor Doug Ducey shut down the work of the experts. Their last "expert" report predicted that the state's peak of cases would not arrive before mid-May.

Time will soon tell if sidelining academic experts in the middle of the "pandemic" was wise or foolish. Potential dangers are obvious when reducing restrictions in any community. Damages are reduced by the fact that a lot of people never followed the restrictions anyway. So there will not be as much of a change. Operating with "real-time" evidence with flexibility to correct is always more reliable than the modeling. The models have been so far off – it's like throwing darts at the board. Donald Trump praised the governors for pulling back on restrictions or non-binding guidelines and hedged his bet by acknowledging *"some people will be affected badly."*

The country needs a turnaround sooner rather than later. The opposition to reopening the country is telling us that you cannot ignore scientific data and bury models. Why? – the projections have consistently been wrong. To me that sounds more like an agenda than a plan.

The petri dish is revealing the culture for all the world to see. We will soon be able to see in real-time the results of the models built by epidemiologists, politicians, or the herd. There are things in play that are coming to light.

A set of detailed documents created by the nation's top disease investigators were written to provide step-by-step guidance to local leaders deciding when and how to reopen the country.

This morning (5/7), the Associated Press (AP) claims to have an "Exclusive" leak provided on the condition of anonymity from within the CDC. The CDC prepared a 17-page *"Guidance for Implementing the Opening Up America Again Framework."* While not authorized to release it, a federal *"informant"* gave a copy to the AP. The "Guide" was designed to control faith leaders, business owners, educators and state and local officials on HOW to reopen the country. The "Guidance" was to be published last Friday, but the White House returned the CDC guidelines as being overly restrictive and in some cases thought it undercut the White House's 3-phase guidelines for opening up the country. The WH guidelines say states should see a 14-day decrease in coronavirus cases before reopening in safe and responsible ways based on their own data and response efforts.

President Trump warned that *"we cannot let the cure be worse than the problem itself."* One size does not fit all. Look at NY and any other region.

TODAY'S WORLD REPORT 05/07/2020 2 PM CT: Covid-19 confirmed 3,813,561 infected worldwide with 267,469 deaths. US confirmed 1,245,622 infected & 75,054 deaths. NY is the epicenter with a reported 26,130 deaths. Reported tests in the US is 7,759,771 and 189,910 Reported recovered.

African Report (5/06) 4,767 new test+ (within ~48 hours) covid-19 cases in 35 countries brings the official count of 56 African countries to 51,254 infected with 1,960 reported covid-19-related deaths. 133 new deaths were reported (within ~48 hours). The continent remains a tinderbox with leaders fearful of the potential. *As of today (5/06) there are 20 deaths reported (Guinea reported 2 deaths in the last 48 hours) in the malaria-*

*troubled countries of **Guinea (11 deaths - 1,856 test+)**, **Botswana (1 death - 23 test+)**, **Burundi (1 death - 15 test+)**, **Zambia (4 deaths - 146 test+)**, and **Malawi (3 deaths - 43 test+)**. 11 African countries out of the 56 countries report 0 fatalities to covid-19.* Pray for Africa.

Chaos in New York – Key Stone Cops
If it were not so painful – you would have to laugh

Glycoscience NEWS Lesson #88 for 2020

05/08/2020 – **The great people of New York deserve better!**

Little dictators across the land are acting like children. They are freeing murders and rapists to make room in the jails for good people who cut hair, stand too close together or don't have on a mask. What judges are doing in Houston is unconstitutional. Tesla was just told workers may not go back to work in San Francisco. The insane control the asylum.

Gov. Andrew Cuomo's office has made serious blunders as they operate out of panic or something. NY took desperate measures to buy medical supplies and equipment **(1)** because they did not stockpile as advised and **(2)** blamed Trump for not delivering supplies... **(3)** which had been delivered and was stored in a NY warehouse, but someone forgot to tell the Governor. **(4)** Early on, Cuomo's office reportedly placed an order for 1,450 ventilators FROM CHINA for $86 million – all cash up front. The contract was canceled but other things are reported to have happened: **(5)** The contract bypassed the state's comptroller because a few days earlier, the governor had issued an emergency order exempting emergency contracts from going through the comptroller.

(6) 80% to 88% of the people died who were put on ventilators -- because of poor oxygen absorption – not ventilator's fault . **(7)** NY had to sign a contract within 4 hours of receiving the quote of ~$59,000 per ventilator and pay the full price of the order up front. **(8)** Wells Fargo froze funds the state had wired to the broker because the bank found a transaction from his account suspicious. **(9)** State officials were told of possible shipping complications and the ventilators may have to be routed through Israel. **(10)** NY was accused of breach of contract. **(11)** Cuomo's team tried unsuccessfully to confirm the stockpile of ventilators in China. **(12)** Scrambling for 40,000 ventilators eschewed the world market with competitive bidding to expedite acquisitions.

(13) Agencies were bidding against one another amid a global shortage... **(14)** causing a ventilator price increase from $25,000 to $45,000. **(15)** Financial incentive for treating covid-19 is on a sliding scale with patient cost estimate ranging from $13,000 for covid-19 treatment without ventilator to $39,000 when placed on a ventilator. **(16)** It is reported that when Wells Fargo froze funds amounting to $59 million after $10 million had already slipped through. **(17)** Lawyers were brought in to figure out who gets the $10 million. My bet is on the lawyers. **(18)** But wait! Wasn't that $86 million to start with? What happened to the other $17 million?

Volunteers went to help NY in the crisis. Some volunteers received a stipend – State Treasury demanded volunteers give them part of their stipend. Oh, there will be lots of stories of many more deaths. Billions of dollars misappropriated. We will discover most deaths were not by covid-19. Congress will write a law that no one can be sued. To view the culture of New York in the petri dish makes one sick. If I were there, I would leave quickly. I love the people but fear the leadership.

Glycoscience advances. More patients call the shots. We win. Forensic investigation exposes the culture of such petrification that it would make anyone angry – but be at peace, a new day is almost here – but it may get worse before it gets better.

Forensic science finds truth that shows what works best and provides the way to cut through barnyard fertilizer, cast aside emotion and biases. Smothered in self-evidence, truth is delivered on a platter so we can make rightful decisions. When findings are complete, those who deceive US will hang on gallows of their own design.

African Update – African Report (5/08) 6,141 new (within ~48 hours) covid-19 positive tests in 40 countries bring the official count of 56 African countries to 57,395 infected with 2,136 reported covid-19-related deaths. 176 new deaths were reported (within ~48 hours). Africa remains a tinderbox with leaders fearful of the potential. As of today (5/08) there are 19 deaths reported in the malaria troubled countries of **Guinea, Botswana, Burundi, Zambia, and Malawi** where quinine may be preventing the spread. 10 African countries out of the 56 countries report 0 fatalities to covid-19. There were 21 new positive cases in **Zambia** reporting a total of 167 with 9 death (no increase last 48 hours). **Malawi** reported 0 additional positive case for a total of 43 also with 0 deaths (no increase).

TODAY'S WORLD REPORT 05/08/2020: Covid-19 confirmed 3,938,080 infected worldwide with 274,898 deaths. US confirmed 1,283,929 infected & 77,180 deaths. NY is the epicenter with a reported 26,243 deaths. Reported tests conducted in the US is 8,408,788 and 198,993 recovered.

"We Don't Have to Have a Vaccine!"

Against the mantra of the WHO, CDC, FDA, AMA, NIH and his own Team

Donald Trump

Glycoscience NEWS Lesson #89 for 2020

05/09/2020 – **We are at war with an invisible enemy.**

We are at war with an invisible enemy that has attacked the human race. The enemy has knocked the whole world to her knees. The petri dish of forensic science allows us to have clear visibility of the damage caused by the invisible enemy. Not subject to any conspiracy theory, a bright light eradicates little tyrants as they dissolve into nothingness.

Let me make it clear that I am not anti-vaccine.
Going against the mantra of the WHO, CDC, FDA, AMA, NIH and his own Team – Donald Trump just told a shattering truth that is self-evident. I am strongly pro-choice – pro-immunization. We need immunization and if a vaccine provides it, great.

The President's 7 words are contrary to the drum beat of Bill Gates and Big Pharma have spent Billions in conditioning the world for a vaccination program for The Big Shot that includes every human being on the planet. These organizations have many outstanding professionals within them but WHO, CDC, FDA, AMA, and NIH are operating in an archaic science enwrapped in bias and prejudice that does not allow for good science to integrate with allopathic medicine. These organizations need the good people within them but the toxic culture needs to be changed. We need to update the culture of each organization to embrace the ideals of the father of medicine – DO NO HARM.

Questioned in a White House briefing if we would be able to return to a normal country before a vaccine is available for everyone, Dr Fauci said that we would not. But, viruses have come and gone – vaccines were not found and the world goes on.

Never let a crisis go to waste

Ebola was to be the earlier vehicle which Big Pharma wanted to complete their plan. I explain this in my book, EBOLA, written during that time. There is indication that the Ebola virus went away because the threshold for herd immunity was achieved. Herd immunity is a mystery that needs to be solved, published, and taught. When the unknown threshold is met within a population, transmission is blocked and the people are safe. No drugs required.

The Current Vehicle to Reshape the World

The coronavirus aka covid-19 aka SARS-CoV-2 is the current vehicle that could reshape the world in the image of the toxic culture of medicine if we let it. The patient can take charge of his and her own health. The cooker-cutter restrictions for New York does not fit Wyoming.

The secret of how to extinguish the coronavirus will be discovered when the threshold is met. Individuals and communities – ranging from families to towns – can hasten the threshold in relationship to their immune resilience.

Forensic investigation is exposing the culture of such petrification that it would make anyone angry – but a new day is almost here – but it may get worse before it gets better.

African Update – (5/09) 3,671 new (within ~24 hours) covid-19 positive tests in 38 countries bring the official count of 56 African countries to 61,066 infected with 2,217 reported covid-19-related deaths. 81 new deaths were reported (within ~24 hours). Africa remains a tinderbox with leaders fearful of the potential. As of today (5/09) there are 25 deaths reported in the malaria troubled countries of **Guinea, Botswana, Burundi, Zambia, and Malawi** where quinine may be preventing the spread. 10 African countries out of the 56 countries report 0 fatalities to covid-19. There were 85 new positive cases in **Zambia** reporting a total of 252 with 7 death. **Malawi** reported 13 additional positive case for a total of 56 also with 0 deaths (no increase). **Pray for Africa**.

TODAY'S WORLD REPORT 05/09/2020: Covid-19 confirmed 4,024,737 infected worldwide with 279,313 deaths. US confirmed 1,309,541 infected & 78,794 deaths. NY is the epicenter with a reported 26,612 deaths. Reported tests conducted in the US is 8,709,630 and 212,534 recovered.

Forensic science finds truth that shows what works best and provides the way to cut through barnyard fertilizer, cast aside emotion and biases.

Coronavirus Winding Down or Heating Up?

Glycoscience NEWS Lesson #90 for 2020

05/10/2020 – COCOA BEACH

The Culture Rises

The scum, the filth, the debris can be clearly observed in the petri dish following certain actions or reactions in the environment – the culture is revealed.

Enjoying the sun and feeling somewhat released from coronavirus lockdown, beach goers on Cocoa Beach, Florida left behind some 13,000 pounds of trash. Few paid any attention to the warning signs "$250 Fine for Littering." One joker said he thought the sign meant that "*it is fine for littering*." The culture of today is too often, "*I don't care*." What remains following an action or a reaction reveals the culture of those involved.

I remember the first *Feast of Joy* we held in front of Houston City Hall in 1990 – we prepared a banquet for a thousand homeless people, cut their hair, gave free clothing, "made their day," encouraged them to advance themselves. My message to the crowd was about our purpose in life – to help others and leave this a better place than we found it. I challenged the homeless and everyone attending to not leave one piece of trash in the park. It was a joyful time of music and dance and feasting. My instructions were to pick up any and every piece of trash and put it in the trash can. They left that park spotless and many Houston parks over the following 30 years each time we had the *Feast of Joy*. The crowd laughed when I told them that every time I go into a rest room, I leave it cleaner than when I entered and that I got into that habit when I started obeying a sign that said, "CLEAN RESTROOM."

Brief Report of Covid-19 Hardest Hit Related Deaths

Reported for 05/10/2020

New York – 26,641	**Illinois – 3,406**
New Jersey – 9,256	**Connecticut – 2,967**
Massachusetts – 4,979	**California – 2,716**
Michigan – 4,555	**Louisiana – 2,286**
Pennsylvania – 3,806	**Florida – 1,721**

USA – 79,525	**Brazil – 11,123**
UK – 31,930	**Germany – 7,569**
Italy – 30,560	**Iran – 6,640**
Spain – 26,621	**Turkey – 3,786**
France – 26,383	**Russia – 1,915**

It's winding down in some areas. Many areas are zero.

African Update – (5/09) 3,671 new (within ~24 hours) covid-19 positive tests in 38 countries bring the official count of 56 African countries to 61,066 infected with 2,217 reported covid-19-related deaths. 81 new deaths were reported (within ~24 hours). Africa remains a tinderbox with leaders fearful of the potential. As of today (5/09) there are 25 deaths reported in the malaria troubled countries of **Guinea, Botswana, Burundi, Zambia, and Malawi** where quinine may be preventing the spread. 10 African countries out of the 56 countries report 0 fatalities to covid-19. There were 85 new positive cases in **Zambia** reporting a total of 252 with 7 death. **Malawi** reported 13 additional positive case for a total of 56 also with 0 deaths (no increase). **Pray for Africa**.

TODAY'S WORLD REPORT 05/10/2020: Covid-19 confirmed 4,101,482 infected worldwide - 282,694 reported deaths. US confirmed 1,329,203 infected & 79,525 deaths. NY is the epicenter with a reported 26,641 deaths. Reported tests conducted in the US is 8,987,524. 216,169 recovered.

This is the Week to Watch – Unusual Reactions – New Discoveries

Covid-19 is causing strange reactions in the mind and body

Glycoscience NEWS Lesson #91 for 2020

05/11/2020 –

Medical scientists are looking at unusual signs and attempting to figure it all out. One common denominator with this invisible killer is that it takes your breath away. Shortness of breath is why ventilators are so needed – yet 80%+ of people put on a ventilator, die. The ventilator is not the cause. Oxygen supply is getting to the lungs but is not being absorbed into the blood.

The signs and symptoms of having covid-19 range wildly. Perhaps most who have the virus are asymptomatic showing no symptoms at all. Some people have neither shortness of breath nor coughing. Some have a fever and some don't. Some react with vomiting and diarrhea and some don't. Some talk and walk and are completely coherent – some are not. The normal oxygen level in the blood may drop WHICH APPEARS TO BE THE FREQUENT INDICATOR. Tests are appearing less reliable.

An unconfirmed report from Tanzania is that President Magufuli submitted a covid-19 test for a goat and pawpaw fruit and they both came back positive for the virus. He announced that the tests have a technical problem.

Some doctors are calling their observations, "*A strange constellation of symptoms.*" The race is on to know how to treat. Valentin Fuster, physician at Mount Sinai Hospital in New York City was quoted, "*At the beginning, we didn't know what we were dealing with. We were seeing patients dying in front of us. It was all of a sudden, you're in a different ballgame, and you don't know why.*" The football personality of the coronavirus – not knowing which way it will bounce – is more unpredictable than most respiratory viruses. The attack on the lungs is but one symptom. It may also attack anywhere from the brain to the toes. It triggers inflammatory reactions and may cause blood clots.

The WHO lists more than 14,600 papers on covid-19. The CDC has repeatedly altered advice in a frantic attempt to keep up with what it is learning on the fly. They do not know why there are so many disease complications involved with covid-19. One virologist simply stated, "*... this is so new that there's a lot we don't know.*"

Let me attempt to put the "Tipping Point" in perspective

Covid-19 attacks the most vulnerable – the individual with a compromised immune system. When the immune system is weak, covid-19 is a pathogen that is pushing many diseases over the tipping point to failure. Health conditions that are near failure are pushed over the edge by covid-19. This fact prevails especially in the area of pneumonia and respiratory illness. If the heart is weak, it may attack the heart. If muscles are weak, it may contribute to the weakness. If vital organs are weakening – it can push them over the tipping point. We discussed in Lesson #76 published on 4/26/2020 how covid-19 can cause blood clots and infect the blood.

TODAY'S WORLD REPORT 05/11/2020: Covid-19 confirmed 4,177,502 infected worldwide - 286,330 reported deaths. US confirmed 1,347,881 infected & 80,682 deaths. NY is the epicenter with a reported 26,988 deaths. Reported tests conducted in the US is 9,382,235 and 232,733 recovered.

African Update – (5/09) 3,671 new (within ~24 hours) covid-19 positive tests in 38 countries bring the official count of 56 African countries to 61,066 infected with 2,217 reported covid-19-related deaths. 81 new deaths were reported (within ~24 hours). Africa remains a tinderbox with leaders fearful of the potential. As of today (5/09) there are 25 deaths reported in the malaria troubled countries of **Guinea, Botswana, Burundi, Zambia, and Malawi** where quinine may be preventing the spread. 10 African countries out of the 56 countries report 0 fatalities to covid-19. There were 85 new positive cases in **Zambia** reporting a total of 252 with 7 death. **Malawi** reported 13 additional positive case for a total of 56 also with 0 deaths (no increase). **Pray for Africa.**

The Public Begins to See the Magic!

Glycoscience NEWS Lesson #92 for 2020

05/12/2020 – HOUSTON

To Discover Magicians' Secrets has always intrigued me. How did the magician do that slight-of-hand trick and his big magic act on stage? It is amazing to see the professionalism of crafty deception on the stage. The world is the stage, covid-19 is the subject of the act, and you observe.

Slight-of-hand is to perform an event so you see what isn't real, but believe it is real. The deception is to get the viewer to believe strongly in the event. The public is encouraged to believe today's report that nearly 300,000 people worldwide have died from covid-19 and that 4,315,679 people are infected. What if the President tests positive?

When we go back stage and look at the props that hold up the covid-19 image, the deception becomes more clear. Reality is not what the audience imagines. The Lancet (I like that medical journal) published a report on March 9, 2020 from information supplied by WHO and CDC outlining the international daily death toll of 26 diseases. TB was at the top of the list with 3,014 deaths per day. Coronavirus covid-19 is 17th on that list at 56 deaths per day. BUT, if you count all deaths it appears the little invisible killer is much more deadly than it really is.

Every organization that has come to destroy US is in serious trouble themselves. We have witnessed in recent days how the World Health Organization (WHO) has been discredited. Has the Center for Disease Control (CDC) been truthful or correct on anything? We have learned that the National Institutes of Health (NIH) funded and was still funding the Wuhan laboratory until the President stopped the money transfers. Big Pharma has a worldwide vaccine agenda to innoculate every human being on planet earth. Italy had a mandatory vaccination program BEFORE covid-19.

There is always going to be a miracle vaccine in a few months. If that were true, then why has there never been a vaccine developed for West Nile, SARS, Bird flu, Swine flu, Ebola, or Zika? And, does the flu shot help or harm?

Truth and Safety First!
Dr. Fauci says, *"Coronavirus deaths are undercounted."*
He may be right in that many asymtomatic
The opposite is true because every disease known to man, it seems, can be counted as covid-19. The not-so-slight-of-hand magic trick is to change the tag name from pneumonia to covid-19 or from TB to covid-19. Anything else that kills a person, let's just call it covid-19. In New York, heart attacks were called covid-19. Statistics do not lie but statisticians do.

Today's count for worldwide deaths from covid-19 is nearly 300,000. Experts are saying that in reality perhaps 1% of the deaths are actually covid-19. If that's the case that would be maybe 3,000.

Dr Fauci tells us that the seniors outbreak is not yet under control – don't come out of lockdown too quickly. The fact is that more people die in lockdown than when they get out in the sun. California ruled that you can go outside if you exercise. You cannot just relax on the beach, you must keep moving. The people are awakening to the idea of using covid-19 to control the masses.

In Africa, Tanzania's president John Magufuli, who holds a doctorate in chemistry, is denying a spike in the country's coronavirus cases, claiming that positive covid-19 tests are fake. It is reported that President Magufuli had a goat and a pawpaw fruit test positive for covid-19.

As lies and deceit are exposed. It may get ugly and whatever it takes, fear must be put back in the people for them to accept a solid social allopathic healthcare system. But the people want to use what **WORKS BEST**. Integrate that into allopathic.

With more testing comes more positive cases and many more asymptomatic cases will be recorded. Texas is reporting 1,000 new positive cases per day for the past 5 days. Hopefully testing will become more trustworthy and when we strip away the fake death reports, we will see, what? Perhaps 1% is real.

Much more good than bad will come out of this virus event for US.

World Report and African Update – ahead.

The people want to use what WORKS BEST. Integrate that into allopathic.

Texas Sheriff Refuses to Enforce Distancing Edict

Overcoming Economic Fear is becoming more important than virus

Glycoscience NEWS Lesson #93 for 2020

05/14/2020 – One size does not fit all

Fear of reopening vs. fear of not reopening is growing. The people are becoming weary because they do not like being controlled like this. What works in Wyoming will not work in New York. Many counties have not one case of covid-19. To require their businesses close and suffer economic hardship may be needless political control.

The pandemic focus is shifting from a health and medical fear to an economic fear. The attitudes are giving rise to a fear of an economic epidemic. As the fear of covid-19 is overshadowed by fear of economics, it becomes evident that the cure can become more dangerous than the plague. Either choice can trigger a devastating toll. To observe people's reaction is fascinating. As quarantines and lockdowns are lifted and freedom returns, what will be the people's response?

Texas General Land Office orders all Texas beaches to reopen.
Despite the threat of covid-19, Galveston County Sheriff will not attempt to enforce the 6 feet distancing edict this weekend during the Crystal Beach Go Topless Jeep Weekend on Bolivar Peninsula. However, more officers will be present. Last year there were more than 100 arrests but not for distancing.

Besides social distancing, the most important thing any of us can do is prioritize how best to support your immune system. The tiny invisible enemy has taught us how important prevention really is.

More strange reports

Coronavirus patients are producing a wide range of strange blood clots throughout the body causing vital organ failure. Manifesting itself in odd ways is causing serious effects on the ability to breathe. A new twist is somehow associated with immune problems in children. We have known for some time that the most obvious symptoms of infection are respiratory. Shortage of oxygen in the red blood cells seems to make it easier for damage to be done on the lining of the blood vessels and allows for unnatural blood clotting.

Inflammation is the killer

Is there really any other killer other than inflammation (unless it is a traumatic event)? With a weakened immune system, localized inflammation happens in the body. A weakened immune system appears to be causing or at least allowing covid-19 to damage the arteries. Several doctors have reported strokes in younger patients and blood clots in the lungs. Doctors admit covid-19 is very confusing and will require time to understand what is going on.

African Report – (5/14) Within the previous week, it appears there has been considerable amount of testing in most of the African countries for covid-19. Currently there are less than 76,000 positive cases reported in 56 African countries with 3,671 reported covid-19-related deaths. Africa remains a tinderbox with leaders fearful of the potential. There are many less reported deaths than projected. As of today (5/14) there are 23 deaths reported in the malaria troubled countries of **Guinea (11), Botswana (1), Burundi (1), Zambia (7), and Malawi (3)**

where quinine may be preventing the spread. 10 African countries out of the 56 countries report 0 fatalities to covid-19.

Here are 3 possibilities for the low death rate in Africa.

(1) Several countries have used a quinine or quinine based drug(s) that are proven by many doctors to be highly successful against covid-19. (2) Perhaps news of Tanzania's president John Magufuli, who holds a doctorate in chemistry, declared that covid-19 tests are fake. He reportedly proved it when a goat and a pawpaw fruit test positive for covid-19. (3) Perhaps the most important factor is that President John and other leaders across Africa are asking their people to pray and ask God to protect them from this curse. To become involved, go to www.AfricaBlessesUS.com. We established this website before the coronavirus pandemic.

TODAY'S WORLD REPORT – 05/14/2020: Covid-19 confirmed 4,443,793 infected worldwide with 302,462 deaths. US confirmed 1,417,889 infected and 85,906 deaths. NY is the epicenter with a reported 27,641 deaths. Reported tests conducted in the US is 10,341,775. 246,414 recovered.

The Glycoscience Institute teaches the values of **Hippocrates, the Father of Medicine**. His value system started with, **FIRST DO NO HARM** and **LET FOOD BE THY MEDICINE AND MEDICINE BE THY FOOD**. The Hippocratic Oath is no longer widely practiced and it is time to get back to it.

Victory against the virus is through immunology, improvement of cellular communication integrity, and modulation of the immune system. Glycoscience holds the answer to the virus chaos.

Beyond Vaccines – Choices and Possibilities
NOW a VIRAL DISINFECTANT INSIDE the HUMAN BODY

– the President was right

Glycoscience NEWS Lesson #94 for 2020

05/15/2020

Allopathic blinders keep traditional medical experts from seeing what is visible to others. This is Big Pharma's Great Opportunity – will they pass the test given them? Medical professionals without self-serving egos welcome WHAT WORKS by integrating a solution previously unknown to them into their traditional practices. This makes them more successful and the patient recognizes that their doctor may have just saved a life. Solutions are available that can transform the medical culture back to earlier days when DO NO HARM was the battle cry.

The medical experts scoffed and laughed! They are still laughing and continue to make fun at how stupid was Donald Trump. The President asked the doctor in front of the world if we could have an internal disinfectant in the body something like we use to disinfectant surfaces and externally on the body. Echos of laughter still reverberate through the halls of Congress and the media airways.

How can you possibly "disinfect" the inside of the human body? Well, the truth be told, we have choices and possibilities but they are outside allopathic medicine. Glycoscience, especially Quantum Glycoscience, is limitless in possibilities. That's the message in my books and lectures.

NOW a VIRAL DISINFECTANT INSIDE the HUMAN BODY

Today's Lesson appears to be outside Glycoscience – but actually is a fundamental part of Glycoscience – I'll explain in a moment. If you are at death's door, I would use applied Glycoscience that I have seen work when everything traditional medicine has thrown at the illness failed.

The President was right when all the world laughed.

There is a "*disinfectant scrubbing solution*" for the inside of the body. It is a 100-year-old cure for viruses that worked on the Spanish flu. But, if you were dying today, the FDA would not allow your life to be saved using this tried-and-proven solution because it is not a toxic drug. This process is still safely used today discreetly and privately in several countries.

The "solution" to remove viruses from inside the body uses a chemical produced by macrophages of white blood cells. [Side note: This explanation is part of how we can improve the immune system through applied Glycoscience because we can activate the macrophage.]

The illegal solution is intravenous hydrogen peroxide dilution of 0.03%. *The Lancet* validates the hydrogen peroxide therapy during the 1919 Spanish flu period. I agree with Heidi Osterman, Chelation Technician, and appreciate her article: "*Intravenous Hydrogen Peroxide: The 100-Year-Old Cure for Covid-19*" published in Health Freedom NEWS Spring 2020. I love that magazine and appreciate the work of its Editor-in-Chief, Attorney Scott Tips, who serves as President of National Health Federation. Scott Tips represents US on the Codex Alimentarius Committee of the United Nations.

African Report – (5/14) Within the previous week, it appears there has been considerable amount of testing in most of the African countries for covid-19. Currently there are less than 76,000 positive cases reported in 56 African countries with 3,671 reported covid-19-related deaths. Africa remains a tinderbox with leaders fearful of the potential. There are many less reported deaths than projected. As of today (5/14) there are 23 deaths reported in the malaria troubled countries of **Guinea (11), Botswana (1), Burundi (1), Zambia (7), and Malawi (3)** where quinine may be preventing the spread. 10 African countries out of the 56 countries report 0 fatalities to covid-19.

3 possibilities for the low death rate in Africa.
(1) Several countries have used a quinine or quinine-based drug(s) that are proven by many doctors to be highly successful against covid-19. (2) Perhaps news of Tanzania's President John Magufuli, who holds a doctorate in chemistry, declared that covid-19 tests used in Tanzania are fake. He reportedly proved it when a goat and a pawpaw fruit tested positive for covid-19. (3) Perhaps the most important factor is that President John Magufuli and other leaders across Africa are asking their people to pray and ask God to protect them from this curse.

To become involved, go to www.AfricaBlessesUS.com.
We established this website before the coronavirus pandemic

TODAY'S WORLD REPORT – 05/15/2020: Covid-19 confirmed 4,542,347 infected worldwide with 307,666 deaths. US confirmed 1,442,824 infected & 87,530 deaths. NY is the epicenter with a reported 27,878 deaths. Reported tests conducted in the US is 10,720,185 and 250,747 recovered.

Big Pharma is in Big Trouble

Glycoscience NEWS Lesson #95 for 2020

05/17/2020

Many neurological challenges are believed to be caused by viruses. We are seeking help from anyone with any neurological symptom to participate in a wide range Six-Month Neurological Pilot Survey using T/C+ with X5 which we believe destroys viruses and repairs some of the neurological damage caused by the virus. More research is needed.

Doctors and healthcare professionals may assist us in monitoring the patient (not required) during the six-month period to help evaluate the changes. Physicians may participate with their patients and we hope to have cases that have a history of health problems. I will be glad to supply more information to those interested without charge.

In the days ahead we will look closer at the science and non-science used regarding covid-19 and forensic science that cuts through all the barnyard fertilizer. Wonders will be revealed. Stay tuned.

Here are possibilities – Why is covid-19 missing in some areas?

231 areas with ZERO - that is an 0 with the rim knocked off – ZERO.

According to a USA TODAY analysis as of 2 days ago (5/15), 231 counties out of 3,143 had ZERO reported cases of covid-19. **(1)** – Some counties are difficult for the virus to find because they are off the beaten path. That contributed to lack of infection rather than immunity. **(2)** – Some areas have not been tested which is another reason for ZERO confirmed cases. **(3)** – Some areas are mainly rural with more isolated sparse populations. In Texas, we had 34 covid-19-free counties out of 254 counties. **(4)** – of the researcher gathering the data said that some counties being free from covid-19 was just luck. NO, I say – that's not science. There is cause and effect and there is no such thing as luck. I remember as a lad in Elm Branch one room school house. I contemplated the physics of rolling dice. And If I had a mechanical hand rolling the dice EXACTLY the same over and over and over – they would always fall the same every time. (But I digress.) **(5)** – Perhaps some counties that have some testing and is still ZERO – is because of flawed testing. They were positive but tested negative. **(6)** – Perhaps some counties where one or more tested positive was flawed testing because they were indeed – negative. **(7)** – Precaution/vigilant can obviously make a difference. **(8)** – "*Divine intervention*" was the response from some people in ZERO areas. "*We're blessed, ... and know things can change daily, but at this moment in time, we're okay.*"

Big Pharma's in Big Trouble

Can they pass the test? They have not developed a successful vaccine for West Nile, SARS, Bird flu, Swine flu, Ebola, or Zika. What makes US believe they can develop a successful vaccine for the coronavirus? They are in over their heads without Operation Warp Speed – with Operation Warp Speed, it will be interesting to see what they do with the long rope they have been given.

Oh, I know: Alex Azar, Secretary of Health and Human Services today (5/17) said that the administration has set a goal for 300 million doses of a coronavirus vaccine to be available to the American people by the end of the year. **That has been Big Pharm's goal for years** – this is their opportunity. Azar said, "*What we're doing is wringing the inefficiency out of the development process to make the development side faster to get to a safe and effective vaccine. And at the same time, we're going to scale up commercial-sized manufacturing to produce hundreds of millions of doses at risk. They may not pan out. They might not prove to be safe and effective, but we'll have it so we can begin administration right away.*"

When findings are complete, those who deceive US will hang on gallows of their own design. Keep your eyes on the CDC.

African Report – will continue – What do the African leaders think?

TODAY'S WORLD REPORT – 05/17/2020: Covid-19 confirmed 4,710,683 infected worldwide with 315,023 deaths. US confirmed 1,486,423 infected & 89,550 deaths. NY is the epicenter with a reported 28,232 deaths. Reported tests conducted in the US is 11,499,203 and 272,265 recovered.

Each of us are agents of change for life or death.

Africa is Helping US

Strange Wonderful things are happening – Not part of the enemy's plans

Glycoscience NEWS Lesson #96 for 2020

05/18/2020 –

Today Donald Trump announced that he is taking the quinine-based drug hydroxychloroquine that has saved countless lives in Africa. "*I'm taking it – hydroxychloroquine,*" the president said. "*A lot of people are taking it. A lot of frontline workers are taking hydroxychloroquine. ... No, I don't own the company. ... I want the people of this nation to feel good. I don't want them sick. And there's a very good chance that this has an impact.*"

My objective is to see US move beyond allopathic medicine to what works best. Quinine – used in Africa as an antimalarial drug in the 1940s and has been an allopathic treatment for auto-immune diseases such as rheumatoid arthritis and lupus. Glycoscience takes us beyond vaccines and drugs and enable us to DO NO HARM. Cheap drugs – hydroxychloroquine – are not vaccines to control people.

In a recent Lesson, I told about the harmful restrictions a Houston judge was placing on the people. Now, the DA is freeing hardened criminals – rapists, murderers – Houston's murder rate is rocketing with more than 100 so far this year. An 80-year-old lady was just stabbed to death. This is part of a bigger agenda for the control of US. The DA's actions kill.

There is hope as long as there is life. There is hope for people to change. There is hope for organizations to change. There is hope for countries to change. Each of us are agents of change for life or death.

Many African leaders are praying for US because they see US under attack, perhaps more clearly than we do. The African is fearful of US when we are represented by WHO, CDC, FDA, and NIH. This is a serious time for North America (Canada and US) and Africa. There are those who would assist the death angel in destroying the people of both continents. I ask our friends here and in Africa to pray for each other as never before – www.AfricaBlessesUS.com. Here are 2 ways Africa is helping US: (1) They ARE praying for US. (2) They proved that quinine can and has saved countless lives and may have immunity.

There are several reasons why the pandemic will fail as we see lies and deception exposed. A lot of good things are going on behind the scenes. It is very important that we keep our immune systems healthy. Prevention/immunity from the tiny invisible killer is the key. We can be immune to its dangers naturally. Eventually, I believe covid-19 will burn out in relation to our immunity and hopefully soon – herd immunity will kick in. We will talk about that later.

Many doctors claim hydroxychloroquine has been remarkably effective – combined with azithromycin and sometimes zinc – in the treatment of covid-19 patients. Peer-reviewed studies in China, France, South Korea, Algeria published from January through April 20 found more than 90% of 2,333 patients treated with hydroxychloroquine and azithromycin improved clinically. Opponents argue that its use for covid-19 hasn't been vetted by clinical trials. The FDA approved hydroxychloroquine covid-19 emergency use in hospitals to treat. Expect a campaign of blaming hydroxychloroquine for covid-19 deaths.

African Report – (5/18) Within the previous 2 weeks, there has been considerable amount of testing in most of the African countries for covid-19. Currently there are less than 89,000 positive cases reported in 56 African countries with less than 4,000 reported covid-19-related deaths. Africa remains a tinderbox with leaders fearful of the potential. There are many less reported deaths than projected. As of today (5/18) there are 28 deaths reported in the malaria-troubled countries of **Guinea (16), Botswana (1), Burundi (1), Zambia (7), and Malawi (3)** where quinine may be preventing the spread. 8 African countries out of the 56 countries report 0 fatalities; 5 countries with 1; 49 countries with less than 100 to covid-19.

3 possibilities for the low death rate in Africa.
(1) Several countries have used a quinine or quinine-based drug(s) that are proven by many doctors to be highly successful against covid-19. (2) Perhaps news of Tanzania's President John Magufuli, who holds a doctorate in chemistry, declared that covid-19 tests used in Tanzania are fake. He reportedly proved it when a goat and a pawpaw fruit tested positive for covid-19. (3) Perhaps the most important factor is that President John Magufuli and other leaders across Africa are asking their people to pray and ask God to protect them from this curse. To become involved, go to www.AfricaBlessesUS.com.
We established this website before the coronavirus pandemic.

TODAY'S WORLD REPORT – 05/18/2020: Covid-19 confirmed 4,805,005 infected worldwide with 318,481 deaths. US confirmed 1,508,598 infected with 90,353 deaths. NY is the epicenter with a reported 28,339 deaths. Reported tests conducted in the US is 11,834,508 and 283,178 recovered.

CDC and WHO are in Big Trouble

Glycoscience NEWS Lesson #97 for 2020

05/19/2020 – WASHINGTON

The President temporarily suspended payments of $400 million per year to WHO until they clean up their act and become independent from China's control. WHO's chief promised to look into mistakes they have made so the temporary suspension would not become permanent. Meanwhile, it is reported today that China pledged $2 Billion to the WHO.

Integrity of CDC's data on the coronavirus has been lacking. Evidence indicates that the CDC has made very little effort to provide accurate data to the public. And their response has been to blame others. This lack of integrity has greatly diminished their position of authority in the minds of observers. There was lack of coordination between other government agencies and the White House. This information was important from day one awareness of the virus. That was the job of the CDC and they did not do it. At a Senate hearing, CDC Director Robert Redfield described the coronavirus tracking system as "*archaic*."

Communication makes everything happen and the CDC had very inadequate communication. The people have suffered and died because of the actions and lack of actions of the CDC. Private organizations must step up and take up the slack of incompetent government agencies. Johns Hopkins University has done that with their Coronavirus Resource Center which appears to be more accurate than the CDC tracking system.

Lack of knowledge contributed to the CDC performance. Serious mistakes were made due to lack of preparedness. Chaos was compounded as new symptoms manifest and morphed again before they could *"put their finger on the virus."* The CDC sent faulty test kits to state and local labs in February. Contamination exposed CDC inability or incompetence that delayed activity that may have saved thousands of lives. The CDC was recognized as one of the most trusted health agencies in the world. **It just lost its credibility during the last few weeks along with the WHO.**

Failure is one of the attributes of government.

The phenomenal opportunity the WHO and the CDC had were missed. The world needs what they purportedly had. They could have remained front and center had they only had the right template that integrates WHAT WORKS into the allopathic template. They failed because on their head were allopathic blinders. **The private sector with qualified integrity and guidelines can go beyond vaccines and drugs to bring to the world the future of medicine – TODAY.**

African Report – (5/18) Within the previous 2 weeks, there has been considerable amount of testing in most of the African countries for covid-19. Currently there are less than 89,000 positive cases reported in 56 African countries with less than 4,000 reported covid-19-related deaths. Africa remains a tinderbox with leaders fearful of the potential. There are many less reported deaths than projected. **As of today (5/19) there are 30 deaths (up 2 last 24 hours)** reported in the malaria-troubled countries of **Guinea (18), Botswana (1), Burundi (1), Zambia (7), and Malawi (3)** where quinine may be preventing the spread. 8 African countries out of the 56 countries report 0 fatalities; 5 countries with 1; 49 countries with less than 100 to covid-19.

3 possibilities for the low death rate in Africa.

(1) Several countries have used a quinine or quinine-based drug(s) that are proven by many doctors to be highly successful against covid-19. **(2)** Perhaps news of Tanzania's President John Magufuli, who holds a doctorate in chemistry, declared that covid-19 tests used in Tanzania are fake. He reportedly proved it when a goat and a pawpaw fruit tested positive for covid-19. **(3)** Perhaps the most important factor is that President John Magufuli and other leaders across Africa are asking their people to pray and ask God to protect them from this curse. To become involved go to **www.AfricaBlessesUS.com**.

We established this website before the coronavirus pandemic

TODAY'S WORLD REPORT – 05/19/2020: Covid-19 CONFIRMED 4,897,492 infected worldwide with 323,285 deaths. US confirmed 1,528,568 infected with 91,921 deaths. NY is the epicenter with a reported 28,558 deaths. Reported tests in the US is 12,233,987 and 289,392 recovered.

Can Glycoscience Unscramble an Egg?

A Virus – A Vaccine – The Scrambled Egg of Autism

Glycoscience NEWS Lesson #98 for 2020

05/20/2020 – The Scrambled Egg of Autism

Quantum Glycoscience considers endless nano possibilities to resolve complex puzzles. Quantum science is wrapped up in entanglement that cannot be easily untangled – like a scrambled egg.

An enlightening thought awakened me at 3AM this morning. I got up and wrote this Lesson dealing with the mysteries of autism.

I went online to see how many articles on autism I had written in the past – 30 references came up. Several years ago, I attended an international autism medical conference here in Texas. I knew no one there. Expecting to learn from these noted medical professionals, I was disappointed. I went away knowing that Glycoscience holds the key to unscrambling the entanglement better than all the presentations I had observed at the conference. Since that time, I have collaborated with universities and research scientists at their request in several countries. We skirted around autism but observed improvement in bipolar, depression, Alzheimer's, Parkinson's, and other neurological challenges when patients ate certain natural biological Smart Sugars. Neurological diseases involve the misfolding of proteins – the entangled proteins are not only scrambled – they have genetic splices sprinkled in the omelette.

Pathologists and neurotoxicologists may have confirmed the autism trigger – aluminum inside brain cells of autism children. It is an unequivocal fact that aluminum is neurotoxic and highly biologically reactive with the ability to damage essential cellular (neuronal) biochemistry. Neurodevelopmental disorders manifest inflammation, microglial activity and elevated Interleukin 6 (IL-6) gene that can act as both a pro-inflammatory cytokine and an anti-inflammatory myokine.

Flawed gene expression may be directly caused by the levels of aluminum detected inside the cells of the brain. The aluminum adjuvant causes an instant and sporadic immune stimulation. That is the justification for putting aluminum in vaccines. Aluminum to the brain cells is like the sudden appearance of a bear and a lion – Yes, that will stimulate you to run but you don't have a place to run to. The child doesn't know what to do. We need to slay the bear and the lion.

In 2007 a reporter said that Donald Trump told him that he thought vaccines may cause autism in children. I observed via TV the President's youngest son and Netanyahu's son in Israel as they became friends. It was an unusual bonding of two boys who both evidenced symptoms of autism. Autistic children are special treasures of intellect bottled up awaiting release. Communication skills are obstructed while the brain capabilities may be that of a savant. Kim Peek (the real Rain Man) was initially diagnosed autistic. Savants show us advanced brain abilities.

In 2019, I was invited to participate in an International Bio-medical Conference. I prepared a lecture and Power-Point-Presentation entitled **Correcting Flawed Gene Expression**. A series of events prohibited me from going to the UAE conference. My evidence was overwhelming with case studies showing how **Applied Glycoscience** has corrected flawed gene expression by safely detoxing the brain through glycosylation –

improving the quantity and quality of glycans, glycoproteins, and glycolipids on and in human cells. The toxins can be safely flushed out of the brain. To improve the communication of the cells is to improve the communication skills of children and adults who have autism. Specific Smart Sugars found in nature help fold the proteins properly. **YES! Quantum Glycoscience can help unscramble the egg of autism.** More research is needed.

African Report – As of today (5/20) there are only 30 deaths reported in the malaria-troubled countries of **Guinea (18), Botswana (1), Burundi (1), Zambia (7), and Malawi (3)** where quinine may be preventing the spread. 8 African countries out of the 56 countries report 0 fatalities; 5 countries with 1; 49 countries with less than 100 to covid-19. (5/18) There has been considerable testing in most of the African countries for covid-19. ~90,000 positive cases reported in 56 African countries with <4,000 reported covid-19-related deaths. Africa remains a tinderbox with leaders fearful of the potential. All of Africa currently has ~1/10th the deaths of New York and many less than projected.

TODAY'S WORLD REPORT 05/20/2020: Covid-19 conformed ~5 million infected worldwide - ~325,000 reported deaths. US confirmed >1,550,000 infected & >93,000 deaths. NY is the epicenter with a reported <29,000 deaths. Reported tests conducted in the US is <13 million. <300,000 recovered.

Dangers of Wearing a Mask

We have learned so much, seen so much wrong, evidenced lies and deceit, witnessed failure. **But remember: more good than bad is coming out of this virus situation.**

Glycoscience NEWS Lesson #99 for 2020

05/21/2020 –

No wonder people are confused – no agency seems to have the answer. Using only the allopathic template, the WHO, CDC, and NIH have had conflicting instructions within their own ranks. It is self-evident that they have been serving a porridge of ignorance, neglect, or often deceit.

About the Mask Issue – to wear or not to wear?
This should clear-up the issue once and for all (tongue-in-cheek)

You were told the best way to not get the virus is to not be exposed. But, if you are exposed and get the virus and you have a good immune system – it is better you get exposed because you will then be immune. You were told the only solution is a vaccine that we don't have and if we did have a vaccine, it would not be safe.

You were told to wear a mask for your protection. Then you were told that it will not protect you from being exposed – it will only protect others from you, from whom they need no protection unless you have the virus. The mask on your face will most likely help protect others from you if you sneeze and they are more than 3 feet away. When you sneeze without a mask, have them back off a little more than 12 feet. That is if you have the virus. And that is, unless you really want them to be immune for life. After all, it is for their good.

The danger to yourself for wearing a mask is that it does not allow the carbon dioxide to escape your face. Your own carbon dioxide is re-breathed and you receive less oxygen into your lungs and more carbon dioxide back. So, now we know that the mask is not only NOT for your benefit – it can actually harm you. But that is not the only way your mask can hurt you – the mask becomes a potential carrier of the virus. If you have touched a contaminated surface with your hand or glove, then touch your mask – you have just contaminated your mask. Beside, if you have worn the mask for more than a few hours; it is probably quite dirty. A person almost died by disinfecting his mask with Lysol. Today's CDC update report is that their previous warnings about surfaces being dangerous was wrong. And that it is alright to touch surfaces – they are probably safe.

I believe a couple of governors have ordered his and her people to go outside must wear a mask, stay at least 6 feet apart and keep moving – exercise. If you stop, you could get arrested. And "NO!" you cannot relax on the beach in the sun where it is healthy. And, for goodness sakes do not get in the water. Mayor Blasio of New York says that if you get in the water – "We will pull you out of the water!" Well, that settles that. This is like circus clowns. Let's all laugh at their foolishness. They would be really funny if they were not so dangerous.

African Report – **As of today (5/21) there are only 30 deaths** reported in the malaria-troubled countries of **Guinea (18), Botswana (1), Burundi (1), Zambia (7), and Malawi (3)** where quinine may be preventing the spread. 8 African countries out of the 56 countries report 0 fatalities; 5 countries with 1; 49 countries with less than 100 to covid-19. (5/18) There has been considerable testing in most of the African countries for covid-19. ~90,000 positive cases reported in 56 African countries with <4,000 reported covid-19-related deaths. Africa remains a tinderbox with leaders fearful of the potential. All of Africa currently has ~1/10th the deaths of New York and many less than projected.

TODAY'S WORLD REPORT – 05/21/2020: Covid-19 confirmed 5,102,424 infected worldwide, <2 million recovered, 332,924 reported deaths. US confirmed 1,577,147 infected, 94,702 deaths. NY is the epicenter with a reported <29,000 deaths. Reported tests conducted in the US is >13 million. <300,000 recovered.

Let's Observe How Sweden Handles the Coronavirus
Sweden did not Lockdown – Should we Lockdown again if we have the 2nd wave?

Glycoscience NEWS Lesson #100 for 2020

05/22/2020 –

Sweden has adopted a very different covid-19 strategy from other Nordic nations to stop the spread by imposing very light restrictions. They avoided a lockdown and keep most schools, restaurants, salons and bars open and placed strong emphasis on personal responsibility.

A study underway by Sweden's Public Health Agency is to determine the potential herd immunity in the citizens. The strategy was criticized by Swedish researchers. Attempting to create herd immunity had very little support. So, the authorities denied that achieving herd immunity was their goal. The allopathic mind-set is that no community has yet achieved herd immunity and a vaccine will get us to herd immunity quicker. How about focusing on getting the population to improve their immune systems so the virus is of little or no consequence?

The percentage of people in Sweden with antibodies is about that of other countries with lockdown. An epidemiological study in Spain showed that 5% of the citizens developed covid-19 antibodies by mid-May. Michael Osterholm, whom I have mentioned in previous Lessons, estimates that 5% to 15% of the US population has been infected.

We are in a world of ambiguity. The WHO, who's word we question, just announced *"herd immunity is dangerous calculation*." Sweden's Chief Epidemiologist reports that the country has an immunity level between 15% to 20% of the citizens. As in other countries, Sweden has failed to prevent a high number of deaths in nursing homes.

The Media Campaign Against Hydroxychloroquine
Many doctors and studies have evidenced hydroxychloroquine benefits. I predicted in an earlier Lesson that a media campaign would soon develop to mislead the world. Today (5/22), headlines lambast the President for his taking hydroxychloroquine and calling the 10 cent pill a "Game Changer." Statistics do not lie but statisticians do. The study published today in *The Lancet* is of 96,000 hospitalized coronavirus patients on 6 continents said to have a higher risk from taking an antimalarial drug than those who did not. Now, let's walk through this carefully. A similar observation of high

risk hospitalized covid-19 patients showed that between 80% and 88% who were put on ventilators died.

The study as reported by the media indicates that people treated with hydroxychloroquine, or the related drug chloroquine, were more likely to develop a type of irregular heart rhythm, or arrhythmia, that can lead to sudden cardiac death. THIS WAS NOT A CLINICAL STUDY. It is based on a retrospective analysis of medical records, not a controlled study. This study may be as subjective as all the people who were declared killed by covid-19 when in reality it was TB or pneumonia or cardiac failure. **GREAT deception is at play to force a vaccine in US.**

Doctors are beginning to see the benefits and possibilities of quinine and that millions of lives may have been saved in Africa with quinine-based drugs. I just heard Dr Ben Carson this week say, *"NO, the pill is not harmful to the President."*

The President said that we will not lockdown again.

African Report –
As of today (5/22) there are only 31 deaths reported in the malaria-troubled countries of **Guinea (19), Botswana (1), Burundi (1), Zambia (7), and Malawi (3)** where quinine may be preventing the spread. **Still 8 African countries out of the 56 countries report 0 fatalities**; 4 countries with 1; 49 countries with less than 100 to covid-19. There has been testing in most of the African countries for covid-19. 104,411 positive cases reported in all of Africa with <4,000 reported covid-19-related deaths. Africa remains a tinderbox with leaders fearful of the potential. New York has ~9 times more deaths credited to corona-19 than all of Africa.

TODAY'S WORLD REPORT – 05/22/2020: Covid-19 confirmed 5,213,483 infected worldwide with 2,058,124 recovered and 338,225 reported deaths. US confirmed 1,601,434 infected with 96,007 deaths. NY is the epicenter with a reported 28,853 deaths. Reported tests in the US 13,398,624 with 350,135 recovered.

Doctors are beginning to see the benefits and possibilities of quinine and that millions of lives may have been saved in Africa with quinine-based drugs.

Murder on the Orient Express

Are we living during the biggest murder mystery of all time?

Glycoscience NEWS Lesson #101 for 2020

05/23/2020

An Unsolved Murder

A coronavirus pandemic was predicted by several people. A plan was conceived for the perpetrator to wait in the shadows for the opportune time. The opportune time started with an accident in a foreign land. People all over the world started dying. A plot was scripted to deflect the blame to others. The perpetrators would appear, dressed as saviors for all the world to see.

Standing in designated positions on stage, they appear prepared. They were trained in the schools of higher learning understanding the Frankenstein monster they know to be the killer. It was an obvious choice for them to be asked to help solve the crime. There was no Lone Ranger. The killer had to have help. The "heros" organize a posse of accomplices assigned to the job of *"solving the perfect crime."*

The murderer is on an accelerating killing spree, leaving the most vulnerable cast along the side of the road like rag dolls. The posse orders the wounded be strategically placed in nursing homes filled with the more healthy. In New York, 5,800 have died in nursing homes. What looks too much like euthanasia has been widely recognized and criticized. Now, they have requested government pass a new law to exempt them from the retribution they have earned.

It is essential for the number of deaths to grow and that fear sweeps the land. The great idea is to count every possible death and credit the murderer with all those deaths. So, they say, "*Let us take the thousands who die daily from tuberculosis, pneumonia, the flu and hey, might as well count the heart failures – there's a lot of them.*"

Then some doctors found inexpensive preventative solutions that would render the murderer harmless and protect the targets. The posse has to discredit their opposition that could blow their cover. So, they concoct lies and give the lies to the press who were deputized into the posse. But something is going horribly wrong. There is only a fraction of the deaths they expected. They thought Africa would be aflame by now. The antidote that the posse is resisting seems to be working in Africa. The posse cannot swallow this bitter quinine pill.

WHO is the first of the posse to be discredited. CDC has so many conflicting answers and misrepresenting arguments that we cannot trust them. The FDA supports the posse as hospitals and pharmacies are asked to not sell the antidote. And, it was the NIH who funded the project in the land of the Orient. The enemies are exposed. Truth will prevail. The posse will be disbanded. The evil plan is coming unraveled.

The train of commerce and LIFE is getting back on track and with God's help, He will make US better than ever before.

African Report –
As of today (5/23) there are only 31 deaths reported in the malaria-troubled countries of **Guinea (19), Botswana (1), Burundi (1), Zambia (7), and Malawi (3)** where quinine may

be preventing the spread. **Still 8 African countries out of the 56 countries report 0 fatalities**; 4 countries with 1; 49 countries with less than 100 to covid-19. There has been testing in most of the African countries for covid-19. 104,411 positive cases reported in all of Africa with <4,000 reported covid-19-related deaths. Africa remains a tinderbox with leaders fearful of the potential. New York has ~9 times more deaths credited to corona-19 than all of Africa.

TODAY'S WORLD REPORT – 05/23/2020: Covid-19 confirmed 5,309,698 infected worldwide with 2,112,096 recovered and 361,239 reported deaths. US confirmed 1,622,605 infected with 97,087 deaths. New York is the epicenter with a reported 29,031 deaths. Reported tests in the US 13,784,786 with 361,239 recovered.

Several countries and regions have ZERO deaths from covid-19.

US Relationship with
WHO Terminated Today

**Everything WHO, CDC, FDA, and NIH
have done concerning covid-19 is under question.**

Glycoscience NEWS Lesson #102 for 2020

05/29/2020 – WASHINGTON/UN

The World Health Organization has mishandled the coronavirus, became a puppet for China, misled the world, failed to reform, and ignored reporting obligations. The US was funding WHO more than $400 million per year. When President Trump placed a temporary halt on funding asking them to reform – they refused. Meanwhile, reports are that China committed to sponsor the organization with $1 billion.

During these months of covid-19, WHO and the CDC have offered projections and estimations that have been wrong more often than right. It remains difficult to find figures that we can count on. Strained extrapolations have projected millions of US dying. The wildly wrong numbers sent shock waves through the population. To keep the momentum going, they threw into the morgues every disease they could think of and called it covid-19.

- - - - - - - - - - - - - - -

**To Lockdown or To Not Lockdown
Another bad decision.**

Some simple FACTS ignored:
1) Covid-19 spreads more easily indoors.
2) Covid-19 is more easily transmitted in closed areas.
3) Covid-19 can be spread through normal AC systems.
4) Risk of covid-19 is lower when you are outside.

5) UV rays from the sun helps destroy the virus in ~10 minutes.
6) Deep breathing outside fresh air is beneficial.
7) Being outdoors has physical and mental health benefits.
8) Exercising outdoors can have more health benefits than indoors.

Hundreds of doctors have asked the President to get us out of lockdown as quickly as possible. The economic and physical disaster from the Lockdown may be more damaging than the virus itself.

A New Concern – Stress and Suicides
This month, a report from doctors in California indicated they have seen more deaths from suicide than coronavirus since lockdowns. Dr Michael deBoisblanc from Walnut Creek, California said on ABC, "The numbers are unprecedented." He added that he's seen a "year's worth of suicides" in the last four weeks alone.

- - - - - - - - - - - - - -

Asymptomatic is Good
Asymptomatic carriers is a good thing because they become immune and contribute to herd immunity. It appears that up to 80% of covid-19 infections are asymptomatic. This statistic was derived from a cruise-ship outbreak when more than 80% of people tested positive for covid-19 but did not have any symptoms. Asymptomatic transmission of covid-19 is common and apparently contributes to the virus "burning out." Out of 217 passengers onboard one cruise ship, 128 tested positive for covid-19 but only 24 showed symptoms. The 104 asymptomatic people is 81% of the positive cases, experienced no symptoms.

To compound the chaos, who knows how many false-positive and false-negative test results are out there with the 16,099,515 tests conducted in the US as of today (05/29/2020).

African Report – We will keep our eyes on Africa

TODAY'S WORLD REPORT – **05/29/2020: Covid-19 confirmed 5,930,035 infected worldwide with 2,495,645 recovered and 365,011 reported deaths. US confirmed 1,747,087 infected with 102,836 deaths. New York is the epicenter with a reported 29,646 deaths. Reported tests in the US 16,099,515 with 406,446 recovered.**

ZERO Coronavirus Deaths in Countries & Regions

**ZERO Coronavirus Deaths in Countries & Regions
How They Kept Infections Down – 19 countries with Zero covid-19-related deaths, 90 countries have under 100 deaths**

Glycoscience NEWS Lesson #103 for 2020

05/30/2020 – Some exciting covid-19 surprises

We are learning much by observing countries and regions where covid-19 did not have the expected impact. Here is today's snapshot of South Korea (11,441 cases - 269 deaths), Taiwan (442 cases - 7 deaths), Hong Kong (1,082 cases 4 Deaths), Vietnam (328 cases - 0 deaths) The 97 million people of Vietnam have reported 0 coronavirus-related deaths as have 18 other countries – a total of 19 countries + Antarctica with ZERO deaths. 90 other countries confirm less than 100 deaths.

Vietnam was successful in overcoming covid-19 even with a lesser advanced healthcare system than others in the region. The country was locked down for 3 weeks but even lifted social distancing rules about a month ago. There have been no reports of infections for more than a month. Schools and businesses are opened and life appears quite normal. Dr. Guy Thwaites who heads the Oxford University Clinical Research Unit in Ho Chi Minh City reported, "I go to the wards every day, I know the cases, I know there has been no death."

Observation indicated that Vietnam saved the country's death toll by REJECTING WHO and making an early preventive response. They began preparing for a coronavirus outbreak weeks before its first case was detected while at the same time WHO maintained that there was no evidence' for human-to-human transmission. Pham Quang Thai, deputy head of the Infection Control Department at the National Institute of Hygiene and Epidemiology in Hanoi is quoted, "We were not only waiting for guidelines from WHO. We used the data we gathered from outside and inside the country to] decide to take action early."

In early January, travelers from Wuhan with a fever were isolated and closely monitored. The government ordered agencies to take "drastic measures" to prevent the disease from spreading into Vietnam. Medical quarantine at border gates, airports and seaports was strengthened. The first 2 covid-19 cases were confirmed on January 23, a father and son (Chinese nationals), who had traveled from Wuhan. The very next day, Vietnam's aviation authorities canceled all flights to and from Wuhan. Prime Minister Nguyen Xuan Phuc on January 27 declared war on the coronavirus. Vietnam had taken action before the WHO declared the coronavirus a public health emergency of international concern. The speed of Vietnam's response was the main reason behind its success.

A few countries, including Vietnam and Taiwan, had valuable life-saving experiences in dealing with the SARS epidemic in 2002 and 2003 plus influenza that prepared them for covid-19.

Georgia (between Turkey and Russia) has developed a new tourist slogan: "A place to take a break from the pandemic." How did Georgia have 757 positive cases of covid-19, 600 recovered, and only 12 deaths?

The country was aggressive but flexible protocols based on safety. They closed schools after it had just 3 confirmed cases. It carefully monitored the spread of covid-19, suspended flights to "hot spots," and initiated health screenings at airports and at the borders. After the first week of March, they quarantined citizens as they returned home in some 7,000 hotel rooms.

Georgia plans to reopen its borders on July 1 but not to everyone. The country is developing what they call "safe corridors." The success in handling covid-19 has opened the door for a surprising public trust in the government. They are looking to renew the tourist trade and may use their old moto, "Our guests come from God."

Earlier this month, more than half of the US counties had no covid-19-related deaths and only 10 states accounted for 70% of all US cases and 77% of all deaths.

African Report – to be continued

TODAY'S WORLD REPORT – 05/30/2020: Covid-19 confirmed 6,057,091 infected worldwide with 2,562,191 recovered and 368,711 deaths. US confirmed 1,769,776 infected with 103,758 deaths. NY is the epicenter with a reported 29,710 deaths. Reported tests in the US 16,495,443 with 416,461 recovered.

We Interrupt this Disaster with Another Disaster

Culture in the Petri Dish is Rotten – Let US Change the Culture

Glycoscience NEWS Lesson #104 for 2020

05/31/2020 – HOUSTON

The Parallel Between the Pandemic and the Riots

Lockdown has increased worry, doubt, fear, anxiety, and deaths that may exceed covid-19 deaths. The pandemic presents a control opportunity to mandate vaccines on every person on the planet in the middle of confusion and chaos fueled by ignorance, neglect, and deception.

The death of George Floyd was the excuse for chaos, rioting, looting, and destruction and had little or nothing to do with George Floyd.

It is reported that paid arsonists and rioters were transported to several major cities including New York, Seattle, Philadelphia, Los Angeles, Houston, St. Paul, Minneapolis, Miami, Chicago, and Brooklin. The Mayor of St. Paul announced that of the arrests made, he thought many if not all were from out-of-town and were there to spark violence. It appears that most of the hirelings are white. Many violent criminals have been released by Houston judges – corrupt judges who replaced good judges because of a well-funded campaign with millions of dollars of outside money earmarked to put the current judges in office.

Nearly 200 arrests were made here in Houston Saturday night with 4 officers injured. Governor Abbott has deployed the Texas National Guard to Dallas, Houston, San Antonio, and Austin.

Some judges have contributed greatly to the violence by freeing the guilty.

Strange things about Floyd's tragic death

A white police officer, identified as Derek Chauvin who held his knee on the neck of George Floyd as he begged for air, was arrested Friday and charged with 3rd degree murder and 2nd degree manslaughter. Third or 4 officers involved should have stopped Chauvin who has a "*checkered background*" with a reported file of 18 complains against him.

George Floyd (46 - 6' 6") grew up in Houston's Third Ward. His body will be returned here. An autopsy report indicates Floyd's cause of death suggest there were a number of contributing factors. The report, conducted by the Hennepin County Medical Examiner, states there were "*no physical findings that support a diagnosis of traumatic asphyxia or strangulation*"; however, the combined effects of being pinned down by an officer as well as "*his underlying health conditions and any potential intoxicants in his system likely contributed to his death.*"

The parallel between the coronavirus and the riots is that they are opportunities. When I wrote the book *Ebola Lies* during that epidemic a few years back, I observed the lies and deceptions about Ebola to do great harm. The same rule book is being used with covid-19. The African race has become more fearful – I have been told – that they believe some people are out to limit their population. Oh, how our culture needs to return to DO NO HARM and BLESS OTHERS!

The parallel of covid-19 and Ebola is that the enemy awaits for a more opportune time to take advantage of the crisis. Ebola seems to have "*burned out.*" Coronavirus aka covid-19 aka SARS-CoV-2 may soon burn out. A vaccine may or may not be developed or work if one is developed. **But this we know – we have learned much the last few weeks:** (1) so new technologies can be developed; (2) that allopathic medicine does not have the answer; (3) that allopathic blinders and that of prejudices and false information cause devastating choices and results.

Devastating choices make for abundant chaos which offers great opportunities to solve problems and bless others. The WHO has been discredited for their devastating choices. The countries that disregarded WHO had fewer deaths. The CDC, FDA, and NIH stood by like passive cops on a murder scene while the murder is taking place. The 4 "protecting" organizations proved conclusively that we need to think outside the allopathic box and integrate Glycoscience and WHAT WORKS into our healthcare. Remember, *"**More good than bad will come out of this situation – if we let it**."*

African Report – As of today (5/31) there are only 36 deaths reported in the malaria troubled countries of **Guinea (23), Botswana (1), Burundi (1), Zambia (7), and Malawi (4)** where quinine may be preventing the spread. **6 African countries out of the 57 countries report 0 fatalities**; 4 countries with 1; 44 countries with less than 100 to covid-19. There has been testing in most of the African countries for covid-19. 147,234 positive cases reported in all of Africa with 4,233 reported covid-19-related deaths. Africa remains a tinderbox with leaders fearful of the potential. New York has >7 times more deaths credited to corona-19 than all of Africa.

TODAY'S WORLD REPORT – 05/31/2020: Covid-19 confirmed 6,164,784 infected worldwide with 2,641,068 recovered and 371,987 deaths. US confirmed 1,789,368 infected with 104,357 deaths. NY is the epicenter with a reported 29,784 deaths. Reported tests in the US 16,936,891 with 444,758 recovered.

BEST NEWS since
SARS-CoV-2 was sent to US

Glycoscience NEWS Lesson #105 for 2020

06/01/2020 – ROME

Reminds me of what a person said a few years ago about a monsoon in Bangladesh, *"This monsoon was a failure – it only killed 100,000 people."* Covid-19 is a failure because today's NEWS is that the virus is losing its potency – it has become less lethal. That is the sentiments of a top Italian doctor.

"In reality, the virus clinically no longer exists in Italy," Reuters quoted Alberto Zangrillo, head of the San Raffaele Hospital in Milan in the northern region of Lombardy – the hot spot of Italy's battle against covid-19. He explained on RAI television that the viral load in quantitative terms is *"ABSOLUTELY INFINITESIMAL COMPARED to the tests] carried out a month or 2 ago."*

Italy has reported the 3rd highest covid-19 death toll – 33,475 with the 6th highest global count of 233,197. 158,355 are reported recovered. Zangrillo explained that some experts were too alarmist about a second wave of infections and politicians need to look at reality. The government remains cautious.

Another doctor from northern Italy told the ANSA news agency, *"The strength the virus had two months ago is not the same strength it has today."* Matteo Bassetti is head of the infectious diseases clinic at the San Martino hospital in Genoa.

As would be expected, WHO almost immediately rebuffed the doctors' statements. It is not over but the flames are dying out.

-274-

More focus will be placed on the economical and physical damage of lockdown.

Herd immunity works in Italy and worldwide. Immunity is the key to winning the battle with the virus – not just covid-19 but all 5,000+ viruses that we must learn to LIVE with or die from.

- - - - - - - - - - - - - - -

Gov. Anthony M. Cuomo of New York
Earns the Uncoveted Petri Dish Award

New York leads the nation in covid-19-related deaths in nursing homes. The governor is responsible. A fraction of his devastating orders include forcing covid-19-infected patients into nursing homes with more healthy people that resulted in more than 5,800 deaths (reported earlier) – figures keep changing. After a few thousand deaths in nursing homes, Cuomo blamed the nursing homes for obeying his orders. Cuomo told the world, *"We've tried everything to keep it out of a nursing home, but it's virtually impossible."* That was a blatant lie to cover euthanasia. He has developed a pattern for ordering people to follow his rules and then blame others for his actions. Does his culture best represent US?

- - - - - - - - - - - - - - -

TODAY'S WORLD REPORT – 6/1/20 6:30PMCT: Covid-19 confirmed 6,246,042 infected worldwide with 2,687,848 recovered and 458,231 deaths. US confirmed 1,808,291 infected with 105,003 deaths. NY was the epicenter with a reported 29,833 deaths. Reported tests in the US 17,340,682 with 458,231 recovered.

African Report – As of yesterday (5/31) there are only 36 deaths reported in the malaria troubled countries of **Guinea (23), Botswana (1), Burundi (1), Zambia (7), and Malawi (4)** where quinine may be preventing the spread. **6 African countries out of the 57 countries report 0 fatalities; 4**

countries with 1; 44 countries with less than 100 to covid-19. There has been testing in most of the African countries for covid-19. 147,234 positive cases reported in all of Africa with 4,233 reported covid-19-related deaths. Africa remains a tinderbox with leaders fearful of the potential. New York has >7 times more deaths credited to corona-19 than all of Africa. To learn more go to **www.AfricaBlessesUS.com**. We established this website before the coronavirus pandemic.

Italy's Lockdown Caused a 30% RISE in Divorce – This report is indicative of man's reaction to the virus

Glycoscience NEWS Lesson #106 for 2020

06/02/2020 –

Relax – do not let lockdown or the pandemic and all of its consequences ruin your day. My wife and I have been together during this time and we have not had arguments any more than normal. This report from Italy about the inceased divorce rate is indicative of the secondary problems caused by the reaction to the virus.

The lockdown and consequences of covid-19 will prove remarkably more deadly than the virus itself. Mental health problems caused by worry, doubt, fear, and anxiety due to the lockdown is becoming evident. Italy blames the 30% increase in divorce filings to the strict national lockdown. Forcing people to stay at home has taken its toll and created a mental health crisis endangering a large number of marriages to be in jeopardy.

Minor things pent up inside can balloon into giant problems that become devastating and impact every factor of life. Anger turns to rage and rage burns cities. Italy has agreed to help solve the divorce problem by letting couples file for divorce by email. Divorces filed online can then have a hearing by video chat between lawyers. Well, guess that solves the problem.

Brazil has now become the epicenter for covid-19
South America's largest country has more confirmed covid-19 cases than any country except US. Latin America's largest nation has reported 555,383 confirmed COVID-19 cases with 31,199 deaths, more than any nation except the US.

- - - - - - - - - - - - - -

TODAY'S WORLD REPORT – 6/2/20: Covid-19 confirmed 6,382,951 infected worldwide with 2,731,340 recovered and 463,868 deaths. US confirmed 1,831,821 infected, 106,181 deaths. NY was the epicenter with a reported 29,968 deaths. Reported tests in the US 17,757,838 with 463,868 recovered.

African Report –
As of yesterday (5/31) there are only 36 deaths reported in the malaria-troubled countries of **Guinea (23), Botswana (1), Burundi (1), Zambia (7), and Malawi (4)** where quinine may be preventing the spread. **6 African countries out of the 57 countries report 0 fatalities**; 4 countries with 1; 44 countries with less than 100 to covid-19. There has been testing in most of the African countries for covid-19. 147,234 positive cases reported in all of Africa with 4,233 reported covid-19-related deaths. Africa remains a tinderbox with leaders fearful of the potential. New York has >7 times more deaths credited to corona-19 than all of Africa. To learn more go to
www.AfricaBlessesUS.com. We established this website before the coronavirus pandemic.

You cannot write a movie script this bad

Today Gov. Anthony Cuomo of New York attacked Bill de Blasio for being a lousy mayor of New York City. Yesterday, we awarded Cuomo the Petri Dish Award and now he has just nominated Blasio for the same award. Blasio may be next. I'm reminded of Mayor Nero (called Emperor) of Rome who allegedly fiddled while Rome burned. But, let us always look at the facts to make sure an event is true. The fiddle / violin was invented in 1500 A.D. and Rome burned for 6 days in July 64 A.D. And records, I am told, indicate that Nero was not in Rome at the time of the fire but he was so disliked by the people that they thought he may have started the fire. It may be difficult for me to wait but the next Petri Dish Award will probably not be given until July.

Plan to Discredit Hydroxychloroquine Backfires
They deceived the 2 most prestigious medical journals Lancet and New England Journal of Medicine

Glycoscience NEWS Lesson #107 for 2020

06/03/2020 –

Immediately after President Trump touted hydroxychloroquine, I stated in a Lesson that "They" will now attempt to discredit the 10 cent pill that has probably saved millions of lives especially in Africa. They will have articles published about how the drug is killing people.

False articles were published that accused hydroxychloroquine of killing people. But then, Lancet and New England Journal of Medicine Published flawed studies that the 10 cent drug was unsafe for treating covid-19. Publishing of these studies triggered a policy change for treating covid-19. And the CDC posted on their website the misleading information – again.

It is unlike these two highly respected medical journals to publish such unsupported research. Especially during a time when so many studies support the opposite. Both journals issued an *"Expression of Concern"* pending investigation.

Brazil is now the epicenter for covid-19
South America's largest country has more confirmed covid-19 cases than any country except US. Latin America's largest nation has reported 584,016 confirmed covis-19 cases with 32,548 deaths, more than any nation except the US.

Covid-19 Hitting UK Hard
More Covid-19 deaths in within last 48 hours in the UK due to a "flare-up" bringing the total to 39,811. Of the 386,073 people who supposedly died of covid-19 worldwide, 39,811 were UK residents. The UK is reportedly second to US on the list of countries with the most covid-19 deaths.

Today's Report on a Few Countries
A study of how each country or region is dealing with covid-19 contains a wealth of information: **Today's Report (06/03/2020) World: 386,073 – US: 107,175 – UK: 39,811 – Brazil: 32,548 – Italy: 33,601 – France: 29,024 – Spain: 27,128 – Germany: 8,602 – Japan: 905 – South Korea: 273 – Israel: 291 – Singapore: 24 – Taiwan: 7.**

TODAY'S WORLD REPORT – 6/3/20: Covid-19 confirmed 6,511,696 infected worldwide with 2,807,420 recovered and 386,073 deaths. US confirmed 1,851,520 infected, 107,175 deaths. NY was the epicenter with a reported 30,019 deaths. Reported tests in the US 18,214,950 with 479,258 recovered.

African Report – As of 5/31 there are only 36 deaths reported in the malaria troubled countries of **Guinea (23), Botswana (1), Burundi (1), Zambia (7), and Malawi (4)** where quinine may be preventing the spread. **6 African countries out of the 57 countries report 0 fatalities**; 4 countries with 1; 44 countries with less than 100 to covid-19. There has been testing in most of the African countries for covid-19. 147,234 positive cases reported in all of Africa with 4,233 reported covid-19-related deaths. Africa remains a tinderbox with leaders fearful of the potential. New York has >7 times more deaths credited to corona-19 than all of Africa. To learn more go to www.AfricaBlessesUS.com. We established this website before the coronavirus pandemic.

Anti-hydroxychloroquine attacks were predicted.

Authors of Anti-Hydroxychloroquine "Study" RETRACT Science Paper in Unheard-of Move

Glycoscience NEWS Lesson #108 for 2020

06/04/2020 –

Science papers published in most respected medical journals were RETRACTED in an unheard of medical reversal.

Today 3 authors of an anti-hydroxychloroquine "study" RETRACTED the study. The authors in a public statement by *The Lancet* announced that they decided to issue the retraction after the drug company, Surgisphere Corp., that provided the research data, refused to share the full, detailed data. Other researchers raised concerns, knowing the "study" was rigged. The authors apologized for *"any embarrassment or inconvenience that this may have caused."*

The Lancet said in a statement that it ***"takes issues of scientific integrity extremely seriously, and there are many outstanding questions about** the drug company] **and the data that were allegedly included in this study."***

Researchers, numbering more than 100, had questioned the data behind the study and about the drug company, which supplied the drug for the study. The fourth author, Dr. Sapan Desai, with the drug company, could not be reached for comment.

- - - - - - - - - - - - - - -

Thank those who protect US!
It is the little things that count so much. Several times each week, I have the opportunities of telling Deputy Sheriffs, Constables, Police, or military personnel, *"Thank you and appreciate your service."* They ALWAYS show appreciation and I just helped make their day.

- - - - - - - - - - - - - - - - -

Today's Report on a Few Countries
A study of how each country or region is dealing with covid-19 contains a wealth of information: **Today's Report (06/04/2020) World: 391,179 – US: 108,211 – UK: 39,987 – Brazil: 34,021 – Italy: 33,689 – France: 29,068 – Spain: 27,133 – Germany: 8,635 – Japan: 905 – South Korea: 273 – Israel: 291 – Singapore: 24 – Taiwan: 7.**

- - - - - - - - - - - - - -

TODAY'S WORLD REPORT – 6/4/20: Covid-19 confirmed 6,635,004 infected worldwide with 2,870,596 recovered and 485,002 deaths. US confirmed 1,872,660 infected with 108,211 deaths. NY was the epicenter with a reported 30,174 deaths. Reported tests in the US 18,680,529 with 485,002 recovered.

African Report – As of 6/04 there are **only 36 deaths** reported in the malaria troubled countries of **Guinea (23), Botswana (1), Burundi (1), Zambia (7), and Malawi (4)** where quinine may be preventing the spread. **4 African countries out of the 55 countries report 0 fatalities**; 5 countries with 1; 43 countries with less than 100 to covid-19. There has been testing in most of the African countries for covid-19. 169,644 positive cases reported in all of Africa with 4,756 reported covid-19-related deaths. Africa remains a tinderbox with leaders fearful of the potential. New York has <7 times more deaths credited to corona-19 than all of Africa.

Impact of Protests and Herd Immunity

Disregarding social distancing and authorities, thousands gather to protest. Thanks, guess we can come out of lockdown now!

Glycoscience NEWS Lesson #109 for 2020

06/12/2020 –

Perhaps it wasn't about the virus after all.

Reversals Galore From February to June
We were told to not congregate as in congregation. We were told that we should stay 6 feet apart or we could be 6 feet under. We were told to lockdown, stay in, self-quarantine or we could die. We were told churches must close but essentials like tattoo parlors could be open.

FEBRUARY, I reported in Lesson #20: **Most Alarming Report yet about Coronavirus – It is Asymptomatic**. Reuters reported that the virus can be spread asymptomatically, that is without symptoms, providing no subjective evidence of existence. **JAMA, (Journal of the American Medical Association)**, published a case study that offers clues about how the coronavirus is spreading and provides evidence that the virus may be very difficult to stop. Another infectious disease expert stated, *"Scientists have been asking if you can have this infection and not be ill? The answer is apparently, yes."* This basically makes all testing worthless!

Asymptomatic carriers were the basis of lockdown and literally destroying the economy of nations.
JUNE, WHO finds 'asymptomatic' carriers not spreading coronavirus. WHO and the CDC reversed themselves about asymptomatic cases. Now, they are not finding secondary transmission onward.

The good news is that asymptomatic spread is not to be a big driver of coronavirus transmission. That changes everything! Do not allow fear to be triggered by the jump in cases which is majorly contributed to more testing. The key to all virus protection is to maintain a healthy immune system. Why have we not seen more emphasis on immunity?

After the dust settles, we will run the numbers on the reduction of deaths from TB, pneumonia, and other diseases. Surely those numbers will have dramatically dropped from the same time last year because even heart attacks were attributed to covid-19. Covid-19 was incentivized to increase the number of deaths.

More good news is that covid-19 appears to be burning out. Is herd immunity on the rise?

Today's Report on a Few Countries
A study of how each country or region is dealing with covid-19 contains a wealth of information: **Today's Death Report (06/12/2020) World: 425,394 – US: 113,672 – UK: 42,566 – Brazil: 41,828 – Italy: 34,223 – France: 29,377 – Spain: 27,136 – Germany: 8,783 – Japan: 924 – South Korea: 277 – Israel: 300 – Singapore: 25 – Taiwan: 7.**

TODAY'S WORLD REPORT – 6/12/20: Covid-19 confirmed 7,632,802 infected worldwide with 3,613,277 recovered and 425,394 deaths. US confirmed 2,049,024 infected with 114,669 deaths. NY was the epicenter with a reported 30,758 deaths. Reported US 547,386 recovered.

African Report – As of 6/12 there are only 37 deaths reported in the malaria-troubled countries of **Guinea (24), Botswana (1), Burundi (1), Zambia (7), and Malawi (4)** where quinine may be preventing the spread. **From all indications it appears that Africa is no longer a tinderbox for the spread of covid-19.** New York has <7 times more deaths credited to corona-19 than all of Africa. To learn more go to **www.AfricaBlessesUS.com**. We established this website before the coronavirus pandemic. We ask African leaders to pray for US.

Are Protests Responsible for Increased Covid-19 Cases?

Glycoscience NEWS Lesson #110 for 2020

06/15/2020 – **Protests and Herd Immunity**

Increased number of covid-19 tests accounts for increased positive cases. As of today (6/15) the number of tests in the US has reached 23,984,592; 2,114,026 positive cases; 576,334 recovered cases; and 116,135 deaths in the US.

Florida has a record rise in covid-19 cases with related deaths nearing 3,000. **Texas** has had a spike in cases with the death count at 2,001. Dr. Dawn Emerick, director of the San Antonio Metropolitan Health District, noted: *"Something is happening in our community. It's not just San Antonio. It is also some of the other larger cities in Texas as well."*

Doctors and healthcare workers have noted that the spike in Texas is beyond the increased reporting of positive test results.

Could the protest gatherings contribute to case increase? It appears to be a combination of factors including relaxed self-control. The first day restaurants were opened for inside dining, people were so excited to be somewhat back to normal, I personally witnessed people casting aside social distancing and hugging as they greeted and departed.

Add covid-19 to the list of 5,000+ viruses that we may have to learn how to live with. Don't count on any effective vaccine. It is up to each of us to take responsibility of our own battle in the virus war and it is better to be offensive than defensive.

Look at DEFENSIVE as stop putting toxins into your body that lowers your immune system. Look at your OFFENSIVE position as taking charge of your health with an aggressive program to properly modulate your immune system. When I was a child, I remember doctors and parents drilled into my brain the thought, *"An ounce of prevention is worth a pound of cure."*

Six states have had fewer restrictions than other states. Let us place these 6 states into the *"petri dish to examine the culture"* to see how the results are as of today (6/15).

South Dakota had the fewest covid-19 restrictions. This is the only state with no mandate to close bars or restaurants. The state reports 5,600 cases with 75 deaths.

Wyoming Governor Mark Gordon never issued an official edict but on March 25 he asked citizens to *"stay home whenever possible."* His request came with orders to close schools, certain businesses, and prohibited gatherings of 10 or more people. These restrictions were reduced as time has passed to gatherings of 25 people. Businesses open to the public including salons, restaurants, gyms, and movie theaters reopened on June 1. The state of Wyoming has seen 1,060 positive covid-19 cases and 18 deaths.

North Dakota Governor Doug Burgum ordered no official lockdown. He closed some businesses in April, but reopened bars, restaurants, salons, athletic facilities, movie theaters, and more on May 1. North Dakota has reported 3,101 positive cases and 74 related deaths.

Iowa Governor Kim Reynolds refused the advice of the state's healthcare professionals for a lockdown order. However, Reynolds signed an order on April 7 to close nonessential businesses until April 30. On May 1, restaurants, fitness centers, and malls were allowed to reopen at 50 percent capacity once the order expired. Since May 28, bars, salons, and movie theaters have also been able to reopen at 50 percent capacity. Iowa has reported 24,046 positive cases and 658 deaths.

Arkansas Governor Asa Hutchinson resisted a lockdown but some businesses closed for a period of time though. The first post-pandemic concert was held in Fort Smith, Arkansas, on May 18. To-date, the Arkansas Department of Health has reported nearly 10,370 coronavirus cases in the state, which have resulted in over 160 deaths.

Nebraska never issued a statewide lockdown, but restricted certain nonessential businesses for a time. Bars, restaurants, gyms, salons, and wedding venues are all allowed to run at 50 percent capacity, and general gatherings of fewer than 25 people were allowed. The state has 16,851 positive coronavirus cases and 220 deaths.

FDA Repeals Emergency Authorization of Hydroxychloroquine
FDA, WHO, CDC, NIH United in Battle Against What Works

Glycoscience NEWS Lesson #111 for 2020
06/16/2020 –

"Quinine" – phytochemical from the cinchona tree bark.

Quinine-based drugs similar to hydroxychloroquine may have been responsible for saving millions of lives in Africa and other countries. Hundreds of doctors vouch that hydroxychloroquine is very beneficial in helping their patients recover from covid-19. The FDA has joined WHO, CDC, and NIH in about-face recommendations and misinformation causing more chaos during this pandemic.

Yesterday (6/15) FDA made an abrupt decision to repeal the emergency use authorization anti-malarial drugs chloroquine phosphate and its less toxic metabolite hydroxychloroquine sulfate as treatments for covid-19. Hundreds of doctors have had life-saving results with their patients recovering from covid-19.

Immediately after President Trump advocated use of hydroxychloroquine as a "game changer" for the treatment and recovery of covid-19, I stated that there would be a war of words that Big Pharma would immediately attack the use of this 10 cent pill telling the public it was killing people. THAT HAPPENED. There were lies and deceits published in the form of science papers. The Lancet and the New England Journal of Medicine withdrew studies thought to be fraudulent.

Million of lives may have been saved in Africa and other countries by the use of quinine-based drugs. Now, millions of people may die because of the FDA's decision to block the use of hydroxychloroquine doses in government stockpile. The science team ridiculed the President for giving his public encouragement for hydroxychloroquine. Peter Navarro, the President's trade adviser, called the FDA move a Deep State blindside by bureaucrats that hate the administration more than their concern to save American lives. Mr. Navarro insisted that the FDA would have "blood on its hands" if any of those studies showed hydroxychloroquine was effective.

Hydroxychloroquine and chloroquine have been approved by the FDA. They have confirmed that the drugs are safe. They say that revoking of emergency use authorization in covid-19 does not change earlier approvals of the drugs by the agency. Hydroxychloroquine is still approved as a treatment for select indications of malaria, lupus, and rheumatoid arthritis, while some versions of chloroquine are also approved for the treatment of malaria. *"FDA has determined that the drugs are safe and effective for these uses when used in accordance with their FDA-approved labeling, and patients prescribed these drugs for their approved uses should continue to take them as directed by their healthcare providers,"* so says a statement in the FDA Frequently Asked Questions.

3 possibilities for the low death rate in Africa.
(1) Several countries have used a quinine or quinine-based drug(s) that are proven by many doctors to be highly successful against covid-19. (2) Perhaps news of Tanzania's President John Magufuli, who holds a doctorate in chemistry, declared that covid-19 tests used in Tanzania are fake. He reportedly proved it when a goat and a pawpaw fruit tested positive for covid-19. (3) Perhaps the most important factor is that President John Magufuli and other leaders across Africa are asking their people to pray and ask God to protect them from this curse. To become involved go to www.AfricaBlessesUS.com. We established this website before the coronavirus pandemic.

Espionage Within NIH

FBI Identifies 121 Scientists – 44 Gone

Glycoscience NEWS Lesson #112 for 2020

399 scientists became of "possible concern."

The FBI identified 121 and 44 were called out by their own institutions. The National Institutes of Health acknowledged that dozens of scientists were fired or resigned during the administration's probe into espionage in US laboratories and universities – and the majority was linked to China. The 44 scientists lost their jobs as part of an investigation launched in August 2018 into NIH grant recipients' failure to disclose their relationship to foreign governments. *Science* magazine reported that 93% received hidden funding from the Communist Chinese government institution.

Michael Lauer, extramural research director for NIH, said the probe targeted 189 scientists at 87 institutions – 80% are Asian males who received 285 grants totaling $164 million. According to Lauer, 133 of the researchers failed to disclose receipts of foreign grants.

NIH Director Francis Collins called the revelations *"sobering."* Charles Lieber, chairman of Harvard University's chemistry department, was indicted June 2020 on charges that he lied to the US government about his work for a Chinese technology school while receiving federal research funds. Last month the President barred Chinese students and researchers with ties to the Chinese military from entering the US.

Another glance at the culture in the Petri Dish
A serious question: Is Dr. Anthony Fauci's decision-making process during the coronavirus pandemic rational, incompetent, and insubordinate? In an interview, he reversed course and suddenly said that he was no longer in favor of endless shutdowns. On CNBC, Dr. Fauci continued to back-pedal from his support for indefinite lockdowns as if the light came on when he stated, *"We can't stay locked down for such a considerable period of time that you might do irreparable damage and have unintended consequences including consequences for health. ... And it's for that reason why the guidelines are being put forth so that the states and the cities can start to re-enter and reopen."* Continuing in his final flip flop, he acknowledged that the majority of the country was ready to reopen the economy. *"In general, I think most of the country is doing it in a prudent way."*

The *New York Times* reporter said that the night before his testimony to the Senate committee, Fauci previewed that his message would be that reopening the economy would lead to *"needless suffering and death."* His major point was to convey to the Senate committee the dangers of trying to open the country prematurely. He had also claimed that the coronavirus should prevent schools from opening in the fall.

As the rift continues with the WH, the President commented that Fauci likes to *"play all sides of the equation."* The CNBC flip-flop interview proved that to be a fact.

Brazil has now become the epicenter for covid-19
6/18/2020 – South America's largest country has more confirmed covid-19 cases than any country except US. Reported 955,377 confirmed cases with 46,510 deaths and 524,249 recovered.

- - - - - - - - - - - - - - -

TODAY'S WORLD REPORT – 6/18/20 at 5 PM CT: Covid-19 confirmed 8,421,357 infected worldwide with 4,115,237 recovered and 450,716 reported deaths. US confirmed 2,182,285 infected with 118,279 deaths. NY was the epicenter with a reported 30,967 deaths. Reported tests in the US 24,937,877 with 592,191 recovered.

African Report – As of today (6/18) there are only 47 deaths reported in the malaria-troubled countries of **Guinea (26), Botswana (1), Burundi (1), Zambia (11), and Malawi (8)** where quinine may be preventing the spread. **5 African countries out of the 55 countries report 0 fatalities;** 4 countries with 1 fatality. 45 countries with less than 100 fatalities attributed to covid-19. 272,816 positive cases reported in all of Africa with 7,206 reported covid-19-related deaths. While there has been a much lower count of deaths than expected, Africa remains a tinderbox with leaders fearful of the potential. To learn more go to www.AfricaBlessesUS.com. We established this website before the coronavirus pandemic.

Science vs Anti-Science.
The public becomes aware which is
which and HOW to know the difference.

Fauci: US has Anti-Science Bias

True Science Allows Integrative Approach for Best Results

Glycoscience NEWS Lesson #113 for 2020

06/19/2020 – **The wise physician** welcomes an integrative approach to provide advanced scientific solutions. The wise patient welcomes wisdom from outside the allopathic pill box to be cured, not perpetually treated 'til death.

True scientists are teachable and ever learning what works best. A biased scientist is blinded from truth and calls the new approach "anti-science" and points blame away from his own failings. After WHO, CDC, FDA, and NIH were wrong more times that we can count, you would think wise leaders of science would welcome common sense and explore "uncommon" sense. Biases are built from hidden agendas and big egos.

This week, a frustrated Anthony Fauci, expressed that we (that's US) have an *"anti-science bias"* problem. He said, *"One of the problems we face in the United States is that unfortunately, there is a combination of an anti-science bias that people are – for reasons that sometimes are, you know, inconceivable and not understandable – they just don't believe science and they don't believe authority. ... So when they see someone up in the White House, which has an air of authority to it, who's talking about science, that there are some people who just don't believe that — and that's unfortunate because, you know, science is truth. It's amazing sometimes the denial there is, it's the same thing that gets people who are anti-vaxxers, who don't want people to get vaccinated, even though the data clearly indicate the safety of vaccines,"* he added. *"That's really a problem."*

In this Glycoscience lesson series on covid-19, I have attempted to scientifically and clearly look beyond the archaic allopathic blinders for answers. The future of medicine is science beyond just cut, burn, and poison. Those who seek true science will integrate what works with the allopathic on the march to cures.

I remember the high school band director observing the band on the football field. From the second floor of the school, he watched the marching band as all of the students made a sharp left turn except one student who marched straight ahead all alone. The student jumped back into position with the band as giggles were heard. Later, while the students were still laughing and embarrassed for their *"out of step"* classmate, the director called the lone student out in front of all the other band members to tell him, ***"Son, you were the only one in step. They all should have marched with you. You got an A+ today."***

To deny what works is anti-science and requires bias. To accept what works is science and anti-bias. To call something safe, when it is not, is anti-science and bias. **DO NO HARM** should again be the rule – the medical oath once had meaning. To lie, deceive, and falsify data is an attack on true science and requires bias. To observe and truthfully report facts is true unbiased science. Patients are demanding good doctors who will peer over allopathic blinders to listen and learn what can be. Only hidden agendas block true science.

Blood Type and Covid-19

Is Type O Blood more resistant to covid-19?

If Type A Blood is more at risk, what does blood type have to do with it?

Glycoscience NEWS Lesson #114 for 2020

06/21/2020 –

Scientists may have found that blood type plays a significant role in combating the coronavirus.

More questions than answers have come from coronavirus aka covid-19 aka SARS-Cov-2. But to be sure, scientists have lots of data to work with. To mine that data and to study the petri dish culture presents hidden information to find cures and make better life and death choices. A European study has linked blood type to the risk factor of viral infection.

In a report, the researchers state, *"Our genetic data confirm that blood group O is associated with a risk of acquiring Covid-19 that was lower than that in non-O blood groups, whereas blood group A was associated with a higher risk than non-A blood groups."* The team discovered that people with Type A blood are a higher risk of infection from covid-19 while people with Type O blood have a lower risk.

At the University of Kiel in Germany, Andre Franke, professor of molecular medicine, led a study of more than 1,900 severely ill coronavirus patients in Spain and Italy, and compared them to 2,300 people who were not sick. They mined the data in a genome-wide association study to find variations. The researchers are not positive that blood type is a direct cause for the differences in susceptibility.

Let me attempt to explain a possibility factor for susceptibility. Life is in the blood! There are four basic blood types: A, B, AB, and O. This simplistic explanation of blood types does not cover the + and - factors that involve an additional protein. Each blood type is determined by how precisely the Smart Sugars are arranged on the surface of the cells. The PLACEMENT OF JUST ONE SUGAR DIFFERENTIATES BLOOD TYPE. That one sugar has the power to determine life or death. Type O blood is the universal donor while AB blood is the universal receiver. Transfusion of the wrong blood type can kill you because wrong blood type is recognized as a foreign agent to be killed by your immune system. When a virus (foreign agent) enters the blood, the sugar structure of a healthy immune cell identifies it as an intruder and signals for it to be destroyed.

You may have some 70 trillion cells in your body with each cell having between 800,000 to a million docking spaces on the surface. Each tiny plot of land (for landing) is quite small, like 1 millionth of the surface of a single cell. On each plot can be a tree made of glycoproteins or a bush made of just glycans. Each bush or tree is a glycoform (with transponder antennae to transmit and receive data) with an untold number of sugar molecules. These sugars are not randomly chosen. These different, yet specific, sugar molecules are linked together to form branched bushes and trees. When the cell is fully glycosylated (heavily populated) the immune system is properly modulated and there is no docking port for the virus to easily attach to the cell.

A healthy operating immune system produces some 50 to 100 billion new cells each day and maintains the glycoforms on some 70 trillion cells. The number of new glycoforms produced by a healthy human is estimated to number 80 quadrillion (that is an 8 followed by 16 zeros) per day. Glycoscience may be millions of times more complex than the genome project. **The type of blood is not as important as the quality of the blood. Life is in the blood!**

In the *Glycoscience Whitepaper*, I attempt to explain in more detail (www.GlycoscienceWhitepaper.com) how 9 specific Smart Sugars construct and enhance the glycoforms through glycosylation to make the operating system (OS) responsible for all cell-to-cell communication of the human body.

Life is in the blood!

Masks Violate OSHA Safety Standards*
Research: Reduces Oxygen Level that can Damage the Brain

Glycoscience NEWS Lesson #115 for 2020

06/23/2020 – HOUSTON

For your health, show this to the Judge.
Yesterday (6/22), Harris County Judge Lina Hidalgo reignited her failed mask edict with a twist – business owners must demand all customers 10 years old or older wear face masks or the owner of the business can be fined $1,000. That's **Politics**.

This is **Science FACT**: the biology and the physics of the mask. We are participating in an unprecedented medical and political experiment.

The physics: The virus is too tiny to be blocked by a mask. Masks are made to block bacteria which are some 10 to 100 times larger than a virus. With the mask, you are breathing less oxygen and more carbon dioxide. You may be protecting others very little but not yourself.

Masks do not meet OSHA* (Occupational Safety and Health Administration) standards – depletes the body of oxygen causing adverse health issues. **Physics test:** Spray mist of something from an aerosol can through a mask and you see that most of the mist goes through the mask.

The Biology Medical Literature Review

Several scientific papers confirm that masks are basically a visible form of false security. Repeatedly, the warning words appear in these studies regarding wearing a **surgical mask**: *"does not reduce the risk of contracting a verified illness."* That is for surgical masks. The public is instructed to wear much more inferior masks.

The NIH National Library of Medicine www.PubMed.gov has an abundance of published papers about masks: 64,012 results for MASKS and 8,338 results for SURGICAL MASKS. Here are 3 **Conclusions** I picked at random.

Conclusion #1: Face mask use in healthcare workers has not been demonstrated to provide benefit in terms of cold symptoms or getting colds. A larger study is needed to definitively establish noninferiority of no mask use. Published *Am J Infect Control* - 2009 Jun.

Conclusion #2: N95-masked healthcare workers are significantly more likely to experience headaches. ... not demonstrated to provide benefit in terms of cold symptoms or getting colds.

Conclusion #3: Published *Epidemiology and Infection* - 2010 None of the studies reviewed showed a benefit from wearing a mask, in either healthcare workers or community members in households.

I find not one study that shows personal benefit to wear a mask in public. 25% of the oxygen we breath goes to the brain and, depending on the thickness of the mask, a significant percentage of oxygen reduction is apparent.

Quote from OSHA FAQ, "Cloth face coverings do not constitute personal protective equipment. Surgical masks are not considered to be PPE if they are being used solely to contain the respiratory droplets of the person wearing them (referred to by OSHA as "source control").

True victory against the virus is through immunology, improvement of cellular communication integrity, and modulation of the immune system. Glycoscience holds the answer to the virus chaos.

The Glycoscience Institute teaches the values of **Hippocrates, the Father of Medicine**. His value system stated with, **FIRST DO NO HARM** and **LET FOOD BE THY MEDICINE AND MEDICINE BE THY FOOD**. The Hippocratic Oath is no longer widely practiced and it is time to get back to it.

Read the *Glycoscience Whitepaper* that explains how the immune system is improved by improving cell-to-cell communication (www.GlycoscienceWhitepaper.com).

Disclaimer: No medical claims are made or intended. More research is needed and purchases help support neurological research.

References for this series: Include agencies in Asia, mainland China, Hong Kong, Japan, Australia, US, Russia, Agence France-Presse, msn news, Reuters, online citizens with cameras, WND, the National Guard in each state, Johns Hopkins University, and OSHA
(On OSHA's website they explain the importance of using face masks, calls them "respirators" and then disclaims their ineffectiveness with these words:
* "Cloth face coverings do not constitute personal protective equipment. Surgical masks are not considered to be PPE if they are being used solely to contain the respiratory droplets of the person wearing them (referred to by OSHA as 'source control')."

75% Fewer Covid-19 Deaths in Areas NOT Locked Down

Glycoscience NEWS Lesson #116 for 2020

06/25/2020 –

Is to NOT Lock Down Better?
US states that did not lock down this spring kept the virus under control better and had fewer deaths than most other states.

Covid-19 may be more contagious than the flu virus but not as deadly except for those with low immune systems. If more people are infected with covid-19 is there more immunity within a community? Herd immunity is a good thing. If the count were actual deaths from covid-19 and not the deaths from pre-existing conditions or whatever disease they want to "not-call-it" – we would not have the fearful numbers that have terrified and shut down the planet.

Let's redefine the pandemic.
We need take a hard look at the number of deaths rather than focus on the number of infections. Have you ever been infected with the flu virus? The number of flu infections is right at 100% of the population. Since covid-19 is more infectious than the flu, there is a good probability that at some point a 100% could become infected and thus become (at lease somewhat) immuned.

The Wall Street Journal reveals the per-capita death rate for the coronavirus is 75% lower in states that did not panic into lockdown mode. Many governors were pressured to lock down their states. Again, with more people testing positive for covid-

19, the same governors are facing another round of browbeating to shut down and further erode the economy.

In an analysis, The Sentinel, a Kansas nonprofit, compared 42 states that shut down most of their economies with 8 states that did not. Admittedly, the residents in the 8 states were outdoors more. But, lock down means you shouldn't go outdoors where more oxygen is. Stay in and/or wear a mask. Also, the analysis showed that from May 2019 to 2020 in the 8 states, employment fell by 7.8% while employment plunging 13.2% in the 42 states.

Is the number of "covid-19 deaths" medical or political? Are we dealing with science or anti-science? Science is factually honest. Very little about this worldwide coronavirus pandemic has been factually scientific or factually truthful.

Math and Aftermath
of Texas Covid-19
Battle at High Noon
for all the World to Witness

Glycoscience NEWS Lesson #118 for 2020

07/01/2020 –

Beginning of the Aftermath
Exploring restaurants, I discovered several in our neighborhood have gone completely out of business. Many others have sparse business. If the study reported in *The Washington Post* is correct,

2% of all US businesses or more than 100,000 business have already been destroyed by the coronavirus. Bars have been closed. Churches were poised to reopen but then backed down because of the "surge." Liquor and drugs are resulting in a significant escalation of overdoses and suicides.

The economical toll on commerce is astounding and will compound. Headlines continue to traumatize the public like this one: *"How Texas lost control of the virus"* and *"Closed Again."*

My cell phone ALERT went off indicating Houston was on virus RED ALERT and to shelter in place. Our blusterous Harris County Judge Lina Hidalgo said during a media briefing *"Today, we find ourselves careening toward a catastrophic and unsustainable situation," ... "There is a severe and uncontrolled outbreak of covid-19. Our hospitals are using 100% of their base capacity now, and are having to start relying on surge capacity."*

Four large hospital systems stated that was not true. Dr. Marc Boom, of Houston Methodist, said, *"... we're concerned that there is a level of alarm in the community that is unwarranted right now."*

The public is awakening to the massive deception powered by ignorance, neglect, and willful misrepresentation of facts. Texas Lt. Gov. Dan Patrick hit Anthony Fauci with a solid right punch to his upper left jaw by saying, *"He doesn't know what he's talking about."* The people, and several whole countries are understanding that the WHO, the CDC, the FDA, and the NIH have been deceiving the world like a master magician performs his illusion tricks.

Math

From the start of the infection, blaming the coronavirus for all the deaths is the MATH part of AFTERMATH. Statistics do not lie but statisticians do. False counting of deaths to get astronomical numbers was needed to put fear in the people. It appears that countries ignoring "professional" instructions saved thousands if not millions of lives.

Today's Report on a Few Countries

A study of how each country or region is dealing with covid-19 contains a wealth of information: **Today's Death Report (07/01/2020) World: 515,691 – US: 128,061 – Brazil: 60,632 – UK: 43,991 – Italy: 34,788 – France: 29,846 – Spain: 28,364 – Germany: 8,995 – Japan: 976 – South Korea: 282 – Israel: 322 – Singapore: 26 – Taiwan: 7.**

World Report – 07/01/2020 Covid-19 reported 10,668,410 confirmed cases worldwide, 5,464,430 recovered, 515,694 reported deaths. US confirmed 2,686,249 infected, 128,061 deaths. NY was the epicenter with a reported 32,643 deaths. Reported US recovered 729,994 recovered.

Update on Africa – As of 7/01/2020 there are 81 deaths reported in the malaria-troubled countries of **Guinea (33), Botswana (1), Burundi (1), Zambia (30), and Malawi (16)** where quinine may be preventing the spread. **From all indications it appears that Africa is no longer a tinderbox for the spread of covid-19.** 55 countries of Africa have had about 31% as many deaths credited to covid-19 as New York. To learn more go to www.AfricaBlessesUS.com. We established this website before the coronavirus pandemic. We ask African leaders to pray for US.

> **How we respond or react determines the inevitable change coming at us at an accelerating speed.**

How Shall We Then Respond?

(Taken from Lesson #117)

Much of the world is in pain and disbelief. Loved ones were lost. Grief hovers the earth. Out of the grief and groans can come much good. To groan may be to silently pray when words are not enough.

It is natural to mourn in the uncontrollable darkness where bad things happen. To mourn should not denote hand ringing. More good than bad will come out of any situation when we let it. The pandemic holds many a paradox with light and learning coming out of the darkness.

Your purpose and my purpose is to bless others, to help others. We are to make the situation better. Our tears can unlock doors because we truly care. When a plague comes, so must good come along side. We are to bring healing to the world and a light that the darkness cannot stop.

We can all pitch-in and lend a helping hand to those worse off. Let the pandemic shatter the box that has held back our individual lights. Your smile can lighten a heart that just experienced a terrifying consequence of foolish handling of the invisible killer. Never ever underestimate your sphere of influence – expand it.

During the plagues of some 500 years ago, Martin Luther in 1527 sent a letter to pastors and civic leaders admonishing them to "remain at their posts" and prepare to lay down their lives for their flocks. He said that he would protect himself by "*fumigating to purify the air, give and take medicine, and avoid places and persons where I am not needed in order that I may not abuse myself and that through me others may not be infected and inflamed with the result that I become the cause of their death through my negligence. If God wishes to take me, He will be able to find me. At least I have done what He gave me to do and am responsible neither for my own death nor for the death of others. I shall avoid neither person nor place but feel free to visit and help them.*" (Luther: Letters of Spiritual Counsel, ed. T.G. Tappert - London: SCM Press, 1955, excerpt from a letter of 1527 as reproduced from *GOD AND THE PANDEMIC - A Christian Reflection on the Coronavirus and its Aftermath* by N.T. Wright)

As I was bringing **Coronavirus INVISIBLE KILLER** to a conclusion, attorney and Biblical authority, Mark Lanier, gave me the book, ***GOD AND THE PANDEMIC - A Christian Reflection on the Coronavirus and its Aftermath*** by N.T. Wright. The Archbishop of Canterbury, Justin Welby, called the book a classic and that he read it with *"pleasure, provocation and profit"* and encourages others to read it – so I did.

I have enjoyed learning from N.T. Wright in person and from his writings. He was professor at the University of St Andrews, fellow at Wycliffe Hall, Oxford, Bishop at Durham and Dean of Lichfield and fellow, tutor, and chaplain of Worcester College, Oxford. TIME magazine asked him to write a piece on coronavirus which triggered the book.

How "professionals" handled the plague in Britain caused Tom Wright to mourn and grieve because of the government's interruption of his churches' long history of medical work caring for the sick and hurting – the church in Britain has a track record of medical assistance dating back to the 1700s.

Wright says, *"Jesus does not need church buildings for His work to go forward. Part of the answer to the question, 'Where is God in the pandemic?' must be out there on the front line, suffering and dying to bring healing and hope."*

Suddenly Tom Wright was told by authorities that the church cannot and must not continue their medical work. He must leave it to the secular *"professionals."*

As I read, **GOD AND THE PANDEMIC**, I grieved with Tom Wright because his work and the medical work of his church to help hurting people was ripped from him by *"professionals"* in the governmental National Health Service.

As I conclude *Coronavirus INVISIBLE KILLER*, I hurt for all those who were, are, and will be hurt directly and indirectly by the coronavirus and for those whose pain was compounded by people unaware of their own ignorance and incompetency. I grieve for the hurting millions. And, I hurt with Tom Wright in the UK as the count passes 45,507 deaths accredited to the coronavirus. And I am distressed because the medical "experts" in darkness do not have the answers and people suffer and die.

I quote Tom Wright, *"As Jesus' followers today grieve in prayer at the heart of the world's pain, **new vocations may emerge, both of healing and wisdom and of holding up a mirror to those in power to show what has needed to be done.**"*

So be it!

Closing Thought

In 1930, Albert Einstein and Edwin Hubble met at the 100-inch Mount Wilson telescope. Together they gazed into the heavens with the best technology then available.

The excitement of learning has brought scientists to look inward and outward. Astronomers observe the constellations with the Hubble telescope while biologists observe the inner parts of creation with the Transmission Electron Aberration-Corrected Microscope. Both extreme optic scientists are discovering new wonders of vast worlds, so different, yet so similar.

Today, this advanced optic technology enables us to gaze into the heavens and into the nanospheres to observe the invisibles and their activities. As each extreme comes into sharper and ever sharper focus, can we touch the face of God?

Robert Jastrow (September 7, 1925 - February 8, 2008) - American astronomer and planetary physicist, NASA scientist, populist author and futurist. Robert was an avowed "agnostic," but maintained that if the Big Bang Theory were fact, *"there was a beginning to the Universe,* [and if there were a beginning] *there was also a Creator."*

In 1978, Jastrow wrote a piece for *New York Times Magazine.* In July 1980 a condensed version was published in *Readers Digest* entitled *"Have Astronomers Found God?"* He closed his article with these words:

"For the scientist who has lived by his faith in the power of reason, the story ends like a bad dream. He has scaled the mountains of ignorance; he is about to conquer the highest peak; as he pulls himself over the final rock, he is greeted by a band of theologians who have been sitting there for centuries."

Additional
Sources and References:

[1] *What You Don't Know About Your Immune System
Can Be Deadly – Understand your TWO immune systems!*
http://www.glycoscienceNEWS.com/pdf/Lesson20.pdf

[2] *Improve Your Brain - Expand Your Mind*
http://www.endowmentmed.org/pdf/ExpandYourMindImproveYourBrainTableOf
ContentsCompressed.pdf

[3]   http://www.NewsmaxHealth.com/Health-News/ebola-symptoms-spread-early/2014/10/09/id/599812/#ixzz3 Fy3kS0rE

[4]   http://fusion.net/video/20107/dr-aileen-marty-tells-fusion-what-she-saw-fighting-ebola-in-nigeria/

[5] http://www.who.int/mediacentre/news/statements/2014/nigeria-ends-ebola/en/

[6] Missing in Action - Why the WHO Failed to Stop Ebola
Time magazine Nov. 10, 2014

[7] http://www.wnd.com/2014/10/if-you-want-to-live-ignore-the-cdc/

[8]   http://www.cbn.com/cbnnews/healthscience/2014/October/Doctor-Warns-Test-for-Ebola-Has-One-Fatal-Flaw-/

[9] http://www.wnd.com/2014/10/nyc-has-its-1st-ebola-case/

[10]   http://www.dailymail.co.uk/news/article-2806532/So-s-point-NYC-Ebola-patient-PASSED-new-enhanced-screening-JFK-Airport-fall-victim-days-later.html

[11]   http://www.wnd.com/2014/10/ebola-victims-without-symptoms-could-still-be-contagious/

[12]   http://www.wnd.com/2014/10/who-urges-sneeze-protection-while-cdc-retreats/#U7MEfYlvMK3tBlwS.99

[13]   http://www.scientificamerican.com/article/let-s-talk-about-ebola-survivors-and-sex/?WT.mc_id=SA_HLTH_20141104

[14]   http://www.afro.who.int/en/clusters-a-programmes/dpc/epidemic-a-pandemic-alert-and-response/outbreak-news/3694-ebola-outbreak-in-democratic-republic-of-congo--update-27-september-2012.html

[15]   http://en.mercopress.com/2014/09/12/congo-reports-31-new-ebola-cases-in-the-week-to-9-September

[16] http://radioopensource.org/bausch/

[17] The Hot Zone by Richard Preston - The Most Dangerous Strain page 256

[18] Ibid

[19]   http://www.wnd.com/2014/10/move-over-ebola-real-killer-is-coming/? cat_orig=health

[20]   http://www.washingtonpost.com/national/health-science/the-ominous-math-of-the-ebola-epidemic/2014/10/09/3cad9e76-4fb2-11e4-8c24-487e92bc997b _story.html

[21]   http://www.sfgate.com/news/medical/article/WHO-10-000-new- Ebola -cases-per-week-could-be-seen-5821246.php

[22] Canadian researchers discovered monkeys can catch Ebola from infected pigs without direct contact:
http://www.dailymail.co.uk/sciencetech/article-2233956/Could-Ebola-airborne-Ne w-research-shows-lethal-virus-spread-pigs-monkeys-contact.html

[23] http://www.wnd.com/2014/10/is-protective-gear-inadequate-to-stop- ebola/

[24]   http://www.scientificamerican.com/article/ebola-spread-shows-flaws-in-protective-gear-and-procedures/?WT.mc_id=SA_TECH_20141014
[25]
Ebola contagion in Spain and Germany raises fears for Europe
http://news.yahoo.com/eu-demands-explanation-spain-ebola-case-084859314.html

[26]   http://hosted.ap.org/dynamic/stories/E/EU_SPAIN_EBOLA?SITE= AP&SECTION=HOME&TEMPLATE=DEFAULT&CTIME=2014-10-06-14-41-22

[27]   http://www.wnd.com/2014/10/obama-brings-in-1900-people-from-another-ebola-nation/

[28] http://www.wnd.com/2014/10/panic-hits-home-of-dallas-ebola-victim/

[29]   http://latino.foxnews.com/latino/health/2014/10/03/border-patrol-on-alert-after-71-people-from-hard-hit-ebola-countries-illegally/

[30]   http://www.Newsmax.com/Newsfront/illegals-immigration-amnesty-executive-order/2014/11/04/id/604991/#ixzz3I6mI0Htd

[31]   http://www.americanthinker.com/blog/2014/10/ebola_czar_a_ population_control_zealot.html

[32] http://www.dailycollapsereport.com/health/ebola-jihad-terrorists-sicken- thousands/

[33] http://personalliberty.com/ebola-outbreak-advantageous-globalists/

[34] http://fathersforlife.org/health/population_control.htm

[35]   http://www.washingtonpost.com/blogs/worldviews/wp/2014/10/24/how-ebola-is-fueling-prejudice-against-gays/

[36]   http://www.huffingtonpost.com/g-roger-denson/is-homosexuality-populati_b_784449.html

37   http://coffee2bsmelt.tumblr.com/post/18016171026/kissinger-eugenics-and-depopulation

38   http://www.wnd.com/2014/10/is-something-besides-ebola-driving-ebola-epidemic/#DWPSQgDFRvJd3qS4.99

39   http://www.firstpost.com/world/returning-ebola-medical-workers-should-not-be-quarantined-says-us-cdc-1775539.html

40   http://www.wnd.com/2014/10/cdc-insists-we-know-how-to-stop-ebola-spread/?cat_orig=health

41 Non-structural glycoprotein (sGP) papers:
http://www.sciencedirect.com/science/article/pii/S0006291X04018959

42 http://www.ncbi.nlm.nih.gov/pmc/articles/PMC3094950/

43 http://jvi.asm.org/content/72/8/6442.full

44 http://jvi.asm.org/content/79/4/2413.full

45   http://article.wn.com/view/2014/11/16/Beating_Ebola_Hinged_on_Sipping_a_Gallon_of_Liquid_a_Day/

46   http://www.bloomberg.com/news/2014-11-16/beating-ebola-hinged-on-sipping-a-gallon-of-liquid-a-day.html

47 http://freedomoutpost.com/2014/10/us-funds-child-beheading-freedom-fighters-billions-tax-dollars/#4wzVF0lSikPiEw8c.99

48 http://www.bestofbeck.com/wp/activism/saul-alinskys-12-rules-for- radicals

49 http://www.wnd.com/2014/10/decorated-combat-troops-sent-to-fight- ebola/

50   http://yournewswire.com/johns-hopkins-scientist-reveals-shocking-report-on-flu-vaccines/

51 http://www.wnd.com/2014/10/cdc-lying-to-public-about-ebola-doctor- says/

52 http://www.wnd.com/2014/10/cnn-savage-has-wild-conspiracy- theories-on-ebola/

53 http://www.ncbi.nlm.nih.gov/pubmed/7580307

54   http://www.foxnews.com/us/2012/10/05/rising-career-us-army-officer-matthew-dooley-halted-for-teaching-soldiers-on/

55   http://www.americanthinker.com/blog/2013/04/pentagon_deep-sixes_lt_col_dooley_for_violating_muslim_pc.html

56   http://conservatives4palin.com/2014/05/obama-administration-frees-36000-convicted-criminal-aliens-awaiting-deportation.html

57   http://www.wnd.com/2014/10/obama-plan-to-expose-u-s-to-more-disease/#1tyM3WF12V20djeL.99

[58]   http://www.newsmax.com/Finance/Stockman-Fed-central-banks-market/2014/10/22/id/602402/?ns_mail_uid=827494&ns_mail_job=1591896_102
32014&s=al&dkt_nbr=ogfd7ijl

[59]   http://www.newsmax.com/Finance/Stockman-central-banks-bubble/
2014/11/13/id/607258/?ns_mail_uid=827494&ns_mail_job=1595354_11142014&
s=al&dkt_nbr=u2dtjyn5

[60]   http://snovalleystar.com/2014/05/07/hospital-enlists-robot-for-high-tech-cleaning-chores

[61]   http://www.NewsmaxHealth.com/Health-News/Ebola-virus-Xenex-robot/2014/10/13/id/600298/#ixzz3HBwRjRXs

[62] *The Shocking Sugar REPORT* - www.DiabeticHope.com

[63] Ibid

[64] Trehalose reduces cell stress and is an antidepressant
http://www.ncbi.nlm.nih.gov/pubmed/23644913

[65] http://www.responsibletechnology.org/gmo-dangers/65-health-risks

Additional supportive educational materials are available:

www.GlycoscienceNEWS.com

About ONE SMART SUGAR
www.OneSmartSugar.com/video.html

Glossary

(Categorically arranged – Not in alphabetical order)

Glyco:	Greek for sugar.
Glycoscience:	Science of sugars.
Glycomics:	Science of sugars.
Glycobiology:	Biology of saccharides, sugar chains, and glycans.
Glycosylation:	Process in which a sugar is attached to another molecule, as attached to another sugar or protein for them to become glycoproteins.
Glycan:	Sugar structures assembled one sugar at a time as building blocks to be conjugated to function as a glycan constructed from different sugars designed to usually be linked and bonded to a protein.
Glycoconjugates:	Conjugated glycans (yoked together)
Carbohydrate:	Organic compound that includes sugars, starches, and celluloses, produced by photosynthetic plants and contain only carbon, hydrogen, and oxygen.

Sugar: A carbohydrate of carbon, hydrogen
                          and oxygen. It is estimated that
                          there are 200 various types of
                          sugars found in nature.

Saccharide: Simple sugar molecule.

Monosaccharide: A carbohydrate that does not
                          hydrolyze (maintains molecular
                          structure), as glucose, fructose, or
                          ribose.

Disaccharide: Two monosaccharides bonded
                          together as in maltose, lactose,
                          sucrose (invert sugar with alkaline
                          stability), and trehalose. Each has a
                          unique function determined by the
                          type of bond.

Oligosaccharide: A carbohydrate made up of a few
                          linked monosaccharides.

Polysaccharide: A complex carbohydrate composed
                          of sugar molecules linked into a
                          branched or chain structure.

Homopolysaccharides: Polysaccharides formed from only
                          one type of monosaccharide
                          subdivided into straight-chain
                          and/or
                          branched-chains.

Heteropolysaccharides: Polysaccharides formed from two or more different types of monosaccharides in straight-chain and/or branched-chains.

Glyconutrient: A food or supplement containing bioactive and functional sugars.

Phytonutrient: Phyto is Greek for plant Bioactive plant nutrient Functional sugars are phytonutrients but all phytonutrients are not sugars.

Phytochemical: Functional sugars are phytochemicals but all phytochemicals are not sugars.

Additional reading:

http://EzineArticles.com/?expert=JC_Spencer

Note: For additional education, you are invited to learn from past and present Lessons and Articles which our team is migrating to
http://TexasEndowment.org

Glycoscience 101 from Amazon or our bookstore

http://ToKillARatBOOK.com

http://www.OneSmartSugar.com

Texas Endowment for Medical Research, Inc.
PO Lockbox 73089, Houston, Texas 77273

http://TexasEndowment.org   281-587-2020

Other books by JC Spencer:

Smart Sugars

Sugars that Speak. Why we should listen!

An introduction to Glycoscience. Easy-to-read for the student while packed with new information for the seasoned medical professional, research scientist, and learned professor.

Smart Sugars is available in soft cover, Kindle and Audio

Smart Sugars [Unabridged] Audible Audio Edition

by JC Spencer (Author), Ross Merrick (Narrator)

Listen on your Kindle Fire or with the free Audible app on Apple, Android, and Windows devices.

Cut and paste link below to listen to a video sample of the reading of Smart Sugars
www.SmartSugars.com/audio

or
http://www.amazon.com/Smart-Sugars/dp/B00OI2H6VW/ref=sr_1_sc_1?ie=UTF8&qid=1414528141&sr=8-1-spell&keywords=Smart+Surgars+audio+book

Editorial Reviews on Amazon – *Smart Sugars* is an easy-to-read book about the breakthrough of sugar technology that will change the way we live. The author explains that some 800,000 to one million transmitting antennae called glycans (actually sugar) coat each of our healthy cells like fuzz on a peach. **Smart Sugars** will help us take the focus off of the disease to treat and cure, because of these new discoveries. Tomorrow's doctors will use Glycoscience diagnostics to read our cells to better determine health and what our health will or can be years in advance. Updated 2020 – Available in softback for only $3.77. Purchase from our bookstore or www.Amazon.com and type in **Enjoy Smart Sugars**

Enjoy Smart Sugars

We have conclusive documentation that we can improve brain function with certain sugars.

This book is your guide to help you improve brain function, overcome stress, and become healthier.

You may wish to participate in a pilot survey.

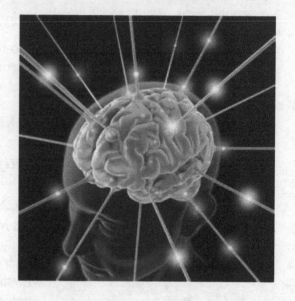

This is not a booklet of ideas. This is an
ACTION book and the next step is up to you.

This Glycoscience textbook is for the learned scientist and inquisitive individual.

Expand Your MIND - Improve Your BRAIN

is an easy-to-read entertaining 580-page science book.
This textbook references over 700 MDs, PhDs, Scientists, Researchers and Educators in the field of Glycomics and Brain Function.

Available Three Volumes in 1 or Volumes 1, 2, and 3 separately.

First Edition published in 2008
Second Edition - 2013 Edition - Updated. Available as an e-textbook, perfect bound 8 1/2 x 11, and hardbound editions

Three Volumes in One $127.77

Vol. 1; 2 and 3 individually as ebook only . $ 27.77 each
      or 3 Volumes in One e-textbook ...... $ 47.77

Order in the Book Store at www.TexasEndowmentmed.org

This is an educational project of The
Endowment for Medical Research

101 Smart Sugars Lessons
that clearly present the progress of
Glycoscience and where it will take us.

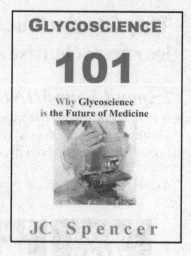

**The Trehalose Handbooks are
available 3 Volumes in 1
or Volume 1, 2 and 3 Separately.**

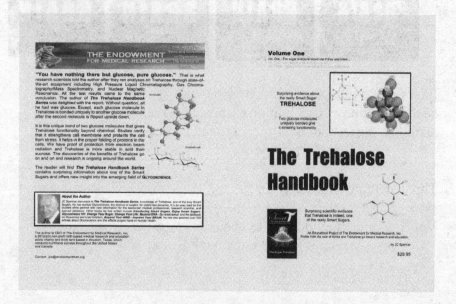

Available in the Book Store at
www.TexasEndowmentmed.org

Video Training for the Healthcare Profession and for the General Public. [This quality of Training would normally cost ~$2,000.]

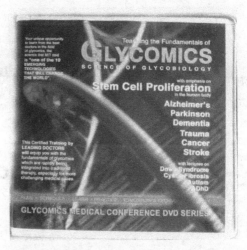

14 hours Professional Glycomics DVD Video Training Series from the Glycomics Conference for Healthcare Professionals. Included is a 500-page syllabus on CD plus all the color slides presented This ADVANCED TRAINING WOULD NORMALLY BE OVER $2,000. Our regular price is $299 -- Readers of *Coronavirus INVISIBLE KILLER* SAVE another $100 off the $299 price $199

14 hours General Public Glycomics DVD Video Training Series from the Glycomics Conference for the General Public (does not include 500-page syllabus of all the color slides presented (SAVE $100 off the $199 price .... $ 99

For details on booking JC Spencer for lectures at universities and fund raising events, contact him at
JCSpencer@TexasEndowment.org

The authoritative Glycoscience whitepaper

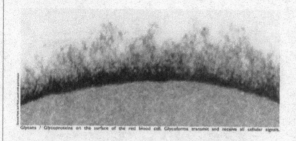

Additional reading and training materials, FREE online support and continuing Glycomics education and research are available.

Readers have access to hundreds of hours of FREE online materials in the form of articles, reports, and video clips. This is a part of the educational effort of Texas Endowment for Medical Research, Inc. and the Glycoscience Institute.

www.GlycoscienceNEWS.com

www.DiabeticHope.com

An educational project of
Texas Endowment for Medical Research
and
Glycoscience Institute

P. O. Box 73089
Houston, Texas 77273

• *Invisible Killer* was loosed on the world. The author looks through the other end of the microscope and examines the culture in the petri dish. Coronavirus – Covid-19 – SARS-CoV-2 are all the same *Invisible Killer* and, indeed, the virus makes a toxic petri dish culture.

• But, that is not the culture *Invisible Killer* observes. The reader begins to observe the culture that brought us the virus that knocked the world to its knees. The author pulls back the covers of deception and chaos.

• Like a master magician, the *Invisible Killer* appeared on the world scene to perform its act as mass murderer. It would take out the most vulnerable people on the planet. Fear would halt earth. The tiny enemy brought us the time the world stood still. When you are invisible, no one can see what you are doing – they can only see the damage you have done. The economic damage, the psychological damage is greater than the virus itself.

• The culture that brought us the *Invisible Killer* becomes more visible as you read. So, the killer mutates and you learn how to follow it. The killer becomes more deadly and you learn how to blunt and then destroy its efforts. You will learn that the only power the virus has over you is that which you give it through a weakened immune system. The virus is the second biggest threat to the human race. The greatest threat to humans is other humans.

JC Spencer is the author of the *Glycoscience Whitepaper* and has written several Glycoscience books. He has studied the works of more than 700 M.D.s, PhDs, Scientists, Researchers, and Educators in the field of Glycoscience and brain function and collaborated / cooperated with schools, universities, and research labs in several countries. He has enjoyed international adventures and speaking engagements in many countries. Since the 1990s, he has worked closely with several specialists, doctors, and healthcare professionals in Glycoscience which resulted in the publishing of peer-reviewed papers evidencing improved brain function in Alzheimer's patients, and pilot surveys for various neurodegenerative challenges including Alzheimer's, Parkinson's, Huntington's, ALS, Lyme, Autism, and ADHD. He passionately envisions the field of Quantum Glycoscience as the proven bull's eye, the Rosetta Stone, the Holy Grail, of medicine and of all healthcare. He is CEO of Texas Endowment for Medical Research, Inc, a 501(c)(3) nonprofit, medical research and education organization and think tank based in Houston, Texas, which conducts nutritional surveys throughout the USA, Canada, and several foreign countries. For more than two decades, he has collaborated with doctors, scientists, and researchers around the world in the fields of brain function, immunology, and Glycoscience.

He is author of the *Glycoscience Whitepaper* - www.Glycosciencewhitepaper.com
For details on booking the author for lectures at universities and fund-raising events
contact him at JCSpencer@TexasEndowmentmed.org

CPSIA information can be obtained
at www.ICGtesting.com
Printed in the USA
LVHW030838040820
661892LV00001BA/7